Metaverse for Beg

Complete Guide on How to Invest in the Metaverse

LEARN ALL ABOUT LAND INVESTING, NFT, AND VIRTUAL REALITY | 5 CRYPTO PROJECTS THAT ARE GOING TO EXPLODE SOON

Table of Contents

Introduction

One of the exciting things about technology is the way it evolves. Sometimes, it goes off on a tangent that surprises everyone and brings about a new reality. However, the next big thing in technology is often predicted years, decades even, in advance. For example, a version of the World Wide Web used currently was predicted many years ago in 1945 by Vannevar Bush.

Bush described something he called the "Memex" as a single device capable of holding all the books you could think of. More than that, he surmises that Memex could also store records, communications and link them together. Talk about a visionary. In his time, he would have been labeled "off his rocker," but his idea is what birthed the "hypertext" some two decades later. The hypertext took another two decades before the World Wide Web, which we enjoy today, was developed.

The Streaming Wars might have just begun, but streaming movies is not a new concept. The very first video was streamed about 25 years ago. This 'out-of-the-blue war between the streaming platforms was hypothesized many years ago. The theory is that the infinite supply of content, the on-demand playback, and the interactivity, dynamic and personalized ads will add to the value of assembling content with distribution.

To have a grand idea of what could happen in the future and see the dreams of icons from many decades ago play out in our present. It also proves that those dream grand dreams in our every day, somewhat sedentary lives; there are those dreaming splendid dreams, making plans, and called fools for technologies that will revolutionize the world in several years. A rough sketch of the future is currently being drawn.

More than that, the ideas are agreed upon, tweaked, and constantly adjusted to fit everyday reality. The technological capacities needed to achieve these dreams are also created daily. However, what is strange is that it is almost unpredictable to guess how these grand ideas will manifest soon. These elegant ideas have already shaped the world in the past.

You can see it in the way books are read now. Rather than going to a bookshop to buy a book, you can simply have one ordered from the Internet. It does not even have to be a physical book; it can easily be an e-book. Even pictures now can be generated digitally, an unthinkable idea in the 1990s. Now, with a simple click of a button, you can have as many images as you would like in as many posts as you can think of.

You can never tell which grand idea would take precedence in the coming future and what innovations will be placed on the back burner. You can tell what competitive dynamic will completely change the sphere of the grand idea and the new experiences that will be produced as a result. No matter how hard one tries to predict the future, life will happen and add its little twist.

Netflix came about with its streaming platform and confirmed to all conversant with the movie-making industry that a change is coming. Even though the handwriting was pretty clear on the wall, time and place were challenged. The idea was to slowly integrate the future into the movie industry seamlessly and efficiently. It took Hollywood almost ten years before the future became the norm.

There are still those in the media industry who fail to grasp what seems to be the basic concept of making money from video game broadcasting and YouTube. The thought of making money by giving real value out for free and charging pennies for optional goodies does not jive with them. It would seem like the transition into the perceived future is difficult to grasp.

The same thing happened to technologists with the concept of a personal computer. Personal computers seem to creep up from out of nowhere that its timing was largely unpredictable. This sudden and fickle change shot Microsoft to the top in an industry where the obvious winner seemed to be IBM. Although Microsoft could read the room clearly, it failed to see the fine print and created a hole that others quickly filled. Its failure to fully understand the importance of operations systems and hardware gave rise to Android and iOS.

Even though Steve Jobs filled a hole created by Microsoft, he did not do a thorough job. Jobs had the right idea at the wrong time and for the faulty device. The Internet was mainly used for instant messaging and email in its early days. You would think there was no possible room for improvement until the late 2000s when the social media networks and platforms cropped up from out of nowhere. Can you imagine life with social media? Although they came late to the party, the concepts used to build these platforms are, shall we say, old science? Everything needed to make Facebook was already pre-Y2K, even though Facebook is an accidental 2005 invention.

When you look at the Internet today, you cannot help but feel proud of everyone and amazed at how far we have come as a people. You also wonder what new and exciting heights we are planning for the future. The next big thing to succeed the Internet has been around the technology community for quite some time now. Since the late 1970s and the early 1980s, the technology community that never seems to stop has started working on something to revolutionize the whole world.

The future state is suspected to be the quasi-successor to the Internet, and it is called the "Metaverse." After conquering the digital world, technology now wishes to extend its tentacles into the physical world. The technological community already has its reach in the physical world with several services and platforms, but now, they are planning on going big.

While the whole idea of the Metaverse is still a tad bit difficult to understand, it is becoming clearer by the day. What seems like a prop from a dystopian future is becoming more and more accurate as the day goes by. The thing about this sort of great, huge, uncontrollable change is that it seems to grow momentum and suddenly sweep everyone off their seats. It is unpredictable, a long game, and will bring a satisfying end to everyone involved.

This means that everyone and their countries have rolled up their sleeves and gotten to work on a new future that is as clear as the stars in the sky. The idea of a Metaverse is bought by many conglomerates who are already working towards bringing that future into a reality. It is the main purpose of Epic Games. It is also the main reason why Facebook bought Oculus VR. After the company's purchase, Facebook soon released its Horizon virtual world/meeting space. This is the first of the many projects that Facebook has in mind. Its many ideas also include AR glasses and brain-to-machine interfaces and communication.

One can only imagine the tens of billions of dollars that will be spent on the research, construction, and experiments of this awesome concept over the coming years. This excellent idea is said to redefine our online-offline virtual future. Soon, Metaverse will be found in the offices of many, if not all, the Big Tech CEOs.

It would have been impossible to explain and even envision what the 2020 internet would be in 1982. It is also almost impossible to fully explain and even understand the Metaverse. And that difficulty is what those who are well versed in the technological world feel. Imagine how much of a head-scratcher it would be for Muggles, the average man.

The grand ideas that can be used to paint a picture would be from science fiction. In these stories, you would imagine the Metaverse to be some sort of digital "jacked-in" internet. It is often depicted as an alternate reality from the physical world. When they hear Metaverse, the idea will be something from the movies Ready Player One and The Matrix in many people's heads. There is no way to tell if that is not what the Metaverse will look like, but it will also be just one facet.

When the Internet first came, and emails could be sent, many people thought it did not get any better than that. The same thought was echoed when fax machines were invented. The same thing was said when people could make long-distance phone calls. The same idea resonated when social media platforms came about. We can all imagine those exact words "it does not get better than this" will also resonate when the Metaverse makes its grand appearance.

The Metaverse's current ideas are just as limiting as Tron depicted the Internet as nothing but a literal digital "information superhighway" of bits. The Internet, as we all know, is much more than that. In the same vein, it would behoove us all to keep in mind that the Metaverse will be much more than what we are currently thinking.

The Metaverse is set to be a digital world that takes bits and pieces from several other worlds to make an all-encompassing world. Social media, online gaming, virtual reality, augmented reality, cryptocurrency, and even the physical world will come together to make the Metaverse. Augmented reality uses visual elements, sounds, and other sensory input to provide the ultimate user experience. On the other hand, virtual reality makes use of complete virtual and fictional realities.

The Metaverse is set to grow and create online spaces where users interact in more multidimensional layers than the current technology supports. Rather than simply viewing the digital content, the users of the Metaverse will find themselves fully immersed in the physical and digital space.

While we cannot entirely give a current and accurate description of what the Metaverse is, some characteristics are believed to be apt for the invention that will soon reshape the way we live our lives. The Metaverse is going to be:

- Be Persistent: Current ideas push the idea that the Metaverse will not have any reason to reset, pause, or end. It is believed that it will run without letting up.

- Be Live and Synchronous: pre-planned events will continue to happen as they do in this current climate. However, the Metaverse will continue to be a living experience of what happens in the physical world for everyone while it is happening.
- Be a Fully Functioning Economy: The Metaverse is set to bring a twist to the way individuals and companies run their businesses. With the quasi-successor, it will become easier to create, own, invest, sell, and be rewarded for an extensive range of work that produces value seen and acknowledged by others.
- Be An Experience: The Metaverse is set to be an experience that fully encompasses the digital and the physical worlds. It takes both the private and the public networks, the open and closed platforms, to produce a unique experience for all.
- Offer Unprecedented Interoperability: with the Metaverse, you can make use of data, digital items, assets, content, and so on from one platform to another with ease. It allows you to use an object like a gun or a car from one video game to another.

These are nothing but the idea of what the Metaverse will do, although not all the suspected roles are widely agreed upon. There are several concerns about whether the users will have a single avatar or digital identity across all platforms or a different avatar for each forum.

The word "Metaverse" snuck out of the tech community into the common space sometime in October 2021. The term was linked to the rebrand that Facebook is currently undergoing. The name "Metaverse" first materialized in a novel written in 1992 called 'Snow Crash.' The book was written by Neal Stephenson, who had repeatedly said he was just making stuff up to make for a great novel. His book was so good and inspiring that it seems the tech community has taken some of his ideas and description to heart. They then went on to work on creating the future that was once the idea of a writer who thought to reality.

The idea of Metaverse is already seen on some low-key scale in the world we are in today. You can see its existence in some form in the gaming world like "Animal Crossing" and "Roblox." There are also links of the Metaverse with the decentralized digital asset called the blockchain. The Metaverse gained the public's attention thanks to the announcement by Mark Zuckerberg, the CEO of Facebook, as he plans to transition into a Metaverse company.

According to Zuckerberg, when he was rebranding his company, the Metaverse will create a platform where you can do just about anything that you could imagine, from working to learning to playing to building and even hanging out with your friends virtually. It will provide you the opportunity to enjoy new experiences that your phones and computers are unable to deliver. It will give you the ability to teleport from one place to another without you having to step out your door.

The arrival of the pandemic caused a massive shift in the way humans interact and do business. Many people, businesses, and platforms are driven online to exist on a digital platform. The other massive events that occurred within the pandemic also drove home the need to escape from the harsh realities occasionally.

The pandemic shift drove home the need for and importance of the Metaverse. It promises a world where you can learn in a classroom without leaving your house or reducing you and your classmates to a square on a screen. It allows for easier and smoother communication between co-workers. With Metaverse, you won't have to miss your favorite concert with your friends all from the comfort of your room.

This book will delve extensively into the Metaverse and why it is a great project to invest in for the coming years. We will also mention our top 5 metaverse projects and why you should invest in them. Keep reading to find out just what the future holds for you!

WARNING!

This book is not financial advice, nor are the concepts in the book intended to replace any financial analysis. Before making a business decision, please do your research. Thank you!

Chapter 1: What Is the Metaverse All About?

Imagine a world where you can virtually be present at a wedding in San Francisco while you're sipping a cup of coffee in Moscow? Or a society where interacting with people is not limited to just blurry video calls and glitchy computer service.

These are the things the metaverse, the next generation of the web, intends to do. But exactly, what is this metaverse? What sets it apart from the rest of the Internet?

A Metaverse is a place where you can interact with virtual items in real-time and with real-time information. In films and shows like Iron Man, Ready Player One, Upload, and The Feed, you've probably already seen this concept being put into practice.

The Metaverse consists of three distinct components. It is first and foremost a technology that allows digital content to be placed on top of the real-world environment. In a way, this is like augmented reality (AR). Using Pokémon Go as a simplistic example, this technology can be improved in future iterations of a metaverse. It's a mix of the digital and real-world stuff. It also uses a hardware gadget to make the natural environment interactive. Using digital material, users can manipulate and interact with media displayed online. The last point includes information on anything and everything in the real world (such as an area, a shop, or a product), along with information about the user (such as the user's timetable). This data will be gleaned through the Internet and machines trained to pick up on users' habits. There are many examples of devices learning from their users' daily behaviors, like Siri (iOS) and Alexa (Android) (on Amazon).

A user's experience is enhanced by obtaining real-time information instantly and virtually through the device into the physical space. At the same time, data is being collected and applied in the background.

To better grasp the Metaverse, one can extrapolate real-world traits to an entirely virtual setting. A metaverse environment will incorporate elements of the actual world into a virtual one. A virtual London or New York, for example, may provide digital representations of real-life streets and buildings to gamers playing in a virtual gaming environment. In a virtual Apple store, you may look through and purchase digital representations of Apple products that will be shipped to your home.

This would, in many ways, represent a continuation of what we currently know as traditional e-commerce. Companies may construct metaverse worlds that not only duplicate the real-life experience but also enhance it, thanks to advances in visual technology and design capability enabled by sophisticated gaming engines like Unreal or Unity. There may be no crowd outside the Manhattan Apple store's digital counterpart at launch time.

The concept of simulating real-world settings in a virtual one is nothing new. It has been going on for a long time with concepts like second life. On the other hand, contemporary online gaming settings have shifted the Metaverse from the old-school 3D block-based worlds of the turn of the century to new, ever-evolving creative ecosystems.

User-created content is the critical difference between the metaverses of the past and today. Playing online games like Fortnite, Roblox, and Minecraft has altered our perceptions of what it means to be "online." When parents wonder why their children spend so much time in these metaverse worlds, it's not because the games or items are well-designed.

Instead, they argue that the metaverses are so compelling and well-designed in and of themselves. People participate, create, and amuse each other rather than just sitting back and watching others. People are paid to make virtual goods in Minecraft, and entire mini-industries have sprung up around them. Fortnite is a virtual stage for real-world music performers to showcase their talents. Every year, tens of millions of people participate in activities that could only occur in the Metaverse.

How Will the Metaverse Work?

To establish a connection with the Metaverse, one will almost certainly require the usage of a device. These goggles, a camera-equipped head-mounted gadget, or other innovative inventions are all possibilities. While they aren't required to participate in the Metaverse, these gadgets may undoubtedly enhance the experience. Users will interact with virtual items in real life by "wearing" a device that integrates all of its components.

To put this idea into practice, imagine waking up every day, donning your metaverse goggles, and entering the Metaverse.

Think this is all just science fiction? No, not at all.

The Google Glass goggles were initially designed to be used for this purpose. You'll be able to observe and interact with virtual information as you walk down the street.

There is a possibility that you are walking to the railway station, and a virtual notification informs you of significant train delays. After that, you have the option of using a quicker route, such as public transportation or carpooling.

This is just a tentative example of how the Metaverse can work in real life. And if you still doubt it is happening, think again. The revolution has already begun.

Virtual items are presented in front of you, in the actual world, in real-time, and you can interact with them. Think of yourself as Tony Stark.

Using your artificial intelligence (AI) helper, you'll be able to find and see the information you're looking for in the actual world virtually. You can then view, click on, or otherwise interact with the things that appear.

Mobile technology has already made it possible for people to live in an enhanced reality, unnerving as it may sound. Your device is aware of your location and time. Since the Internet was created, integrating the real world with the virtual world has been an ongoing process. The Metaverse's primary function is to provide a means of distancing oneself from the realities of the real world. Adventures and alternate lives are possible in Fortnite for those who wish to do so. This escapist worldview has seen a substantial transformation in recent years with the inclusion of real-life components. Is watching a movie on Roblox something you'd like to do? You're playing Grand Theft Auto, and you'd want to buy some sneakers. On TikTok, you may watch a K-pop band's most recent live performance. As trade and engagement move online and into virtual worlds, the Metaverse is fueled by this virtual and real-life convergence. What are the commercial uses of the Metaverse, and who will reap the benefits? To put it another way, the introduction of the Metaverse will change our lives forever. Every industry has potential for metaverse applications.

The possibilities the Metaverse will unlock are endless, from consumer-driven ones like retail to manufacturing and construction and beyond.

It is possible to make purchases in a flash. You won't even have to touch your smartphone to see a product when you see it in a store or on the Metaverse. Through a single account, customers may buy products and compare pricing. Due to improved connectivity, businesses will sell their products anywhere globally, regardless of where their retail locations may be.

Thanks to this new technology, businesses and celebrities will reach a far broader audience and collaborate more easily. In the future, customers will be able to communicate directly with brands. If employed correctly, this could have an excellent commercial influence. Brands and celebrities will see an increase in exposure. There may even be a market for virtual real estate in the Metaverse. Non-fungible tokens and other digital products and property will be given more attention in the future (NFTs). Because they aren't subject to wear and tear, items that can be traded are more valuable. Players may look forward to more immersive and interconnected game worlds in the future. A skin or item acquired in one game can be utilized in another game or swapped for a different item. As virtual cinema allows for private viewings with friends, the social experience will also shift. Many businesses, including journalism, social media, technology, and retail, will find new monetizing methods. At the same time, people will meet, work, and socialize more agreeably online in the future. Intellectual property will play a significant role when it comes to creative activity. Because of its greater accessibility, information, products, entertainment, and social experiences are likely to benefit consumers the most from the Metaverse. The technology market will be dominated by hardware and software enterprises. Providing hardware and software for the Metaverse is expected to rise significantly. Businesses will be able to build their virtual worlds. This means that more people will see and hear about brands and celebrities. As technology improves, so will the potential to provide customers with more relevant commercial offerings and experiences.

Law and rules in the Metaverse are still in flux. Therefore, legal guidance will also be required. As the virtual and real worlds merge, there is a tremendous demand for assistance in data protection, privacy, advertising rules, and ensuring that commercial firm intellectual property assets are secured. For years to come, attorneys and legislators will face the issue of making sure that real-world laws are appropriately translated into the virtual world. The Metaverse is being built by whom? The gaming industry is one of the best examples of how the Metaverse can be used in business today. We can see how the Metaverse can transform the way people interact with digital and real worlds through games like Fortnite and Roblox. As a result, many of the gaming industry's biggest names are also at the forefront of technological innovation and growth in this sector. Take the game Roblox, for instance. The gaming firm, which went public in March 2021, partially laid out its ambition for the company and the adoption of the Metaverse in its prospectus. For Roblox, a pervasive human co-experience platform is the aim, as computing power, high-bandwidth Internet connections, and human interface technologies continue to increase (and even build an economy based on its currency, Robux). Second Life's founders, Linden Labs, also developed their currency and had a bigger GDP than some small countries at one point.

In this situation, user experience is only one factor. Using the prefix "meta" (meaning beyond) and the stem "verse," the word "metaverse" is formed (meaning the universe). Critics believe that several critical features must exist for the Metaverse to fulfill its full potential, including:

1. Persistence

2. The ability to give live, synchronous experiences
3. Interoperability
4. Value creation.

Many stakeholders (individuals, commercial companies, and governments) are expected to be involved in the creation and operation of the Metaverse. This makes sense. Developing a community of stakeholders in the Metaverse, like the current Internet, is necessary for new technologies, businesses, services, content creators, standards and protocols, legislation, and more.

Microsoft's HoloLens augmented reality headset and Facebook's recent purchase of Oculus VR, as well as Unity's significant investment in digital twin technology, all point to the fact that many of the current technology industry giants, such as Microsoft, Facebook, and Unity, will almost certainly play a significant role in the development of the Metaverse.

In the future of the Metaverse, there's no consensus on how it will work, who will develop it, or who will "own" it (if anyone).

However, the broad consensus is that it will exist and no longer be considered a figment of our imaginations.

No matter what happens in the future, one thing is clear:

The Metaverse will expand incrementally over time as capabilities evolve and synergies are formed.

Chapter 2: The Metaverse and Virtual Reality

If you read anything about the Metaverse, you can't help but notice how it resembles virtual reality. However, there are several crucial differences.

Here are some key differences between virtual reality and the Metaverse that you should be aware of if you want to know more.

Unlike Virtual Reality, The Metaverse Isn't Well-Defined

Even though virtual reality is well-understood, the Metaverse is still a bit of a mystery.

Zuckerberg, the CEO of Facebook, describes the Metaverse as "an embodied internet where instead of merely watching the content- you are in it."

Zuckerberg, the CEO of Facebook, describes the Metaverse as "an embodied internet where instead of merely watching the content- you are in it." As opposed to this, Microsoft calls it "a permanent digital environment populated by digital twins of people, places, and objects.".

If we compare these descriptions to what we know about virtual reality, they're incredibly ambiguous. Another possibility is that no one has a definitive definition, not even the IT corporations.

For Facebook, rebranding was an essential aspect of creating a "metaverse." To better reflect their work, they came up with a new name. However, this is by no means the only conceivable explanation. Facebook has a public relations issue.

One may argue that the Metaverse is simply a euphemism for new technological advancements in the current internet infrastructure.

Facebook Does Not Own Either of The Technologies It Uses to Communicate With Its Users

Another dilemma that could arise concerning the Metaverse is who can define it.

The Oculus Rift is owned by Facebook, which significantly impacts virtual reality. But there are several players in the industry, and they're just one.

The Metaverse is no different. Many more companies are collaborating in this rebranding of Facebook as Meta. Microsoft Mesh is a contemporary example of Microsoft's mixed-reality platform, which is comparable to the Metaverse and its many meanings. As a recent Facebook statement clarifies, they see themselves contributing to the Metaverse rather than creating it from the ground up.

This means that the Metaverse will be much larger than anyone corporation can handle, like virtual reality.

There Is a Shared Virtual World in The Metaverse

Users will connect to the Metaverse, a shared virtual area, over the Internet. It should be noted that this functionality is already built into VR headsets.

The Metaverse's virtual realm sounds a lot like the existing virtual reality programs.

Personal avatars will be used to identify and engage in virtual environments. NFTs, for example, will be available for purchase or construction by users.

On the other hand, Metaverse sounds like it will have access to the entire Internet, unlike existing virtual worlds.

Virtual Reality Will Allow Access to The Metaverse

You won't need a VR headset to experience the Metaverse. Many portions of the service are expected to be available through headsets.

Because of this, the border between browsing the Internet and using virtual reality is likely to grow blurred. In the future, virtual reality (VR) headsets may be utilized for tasks currently done on smartphones.

VR may become less of a niche product if Facebook's Metaverse becomes as popular as expected.

VR Will Not Be the Only Means of Accessing The Metaverse

However, as stated in the previous paragraph, the Metaverse will not be limited to virtual reality. As a result, it will be accessible via augmented reality devices and any other device you now use to access the Internet.

When combined with virtual reality, this allows for a wide range of new possibilities. For example, the Metaverse can be brought into the real world using augmented reality.

In addition, virtual worlds will be created to be accessed from any location without a headset.

The Metaverse Has the Potential to Be Much Larger Than The Virtual World Of Virtual Reality

Education, healthcare, and sports are just a few of the uses of virtual reality. However, it is still seen as a form of entertainment.

In terms of scope, Metaverse sounds more like a new and enhanced version of the Internet. In contrast to virtual reality, which many people have ignored completely, the Metaverse is predicted to revolutionize how people work, use social media, and even surf the web.

Is The Internet Going to Be Replaced by The Metaverse?

People had hoped that virtual reality would have a more significant impact on the world than it has so far. Many people aren't willing to use a headset for long periods.

Virtual reality headsets will not be required to access the Metaverse, which will be accessible to both people who have and those who don't. As a result, some people believe it will have a considerably more significant influence.

The Metaverse, on the other hand, is unlikely to replace the Internet completely. An alternative to computer screens can be found in virtual reality headsets. If you're tired of the Internet, the Metaverse can be a fun change of pace. However, neither one is meant to serve as a substitute.

Chapter 3: Metaverse and NFT Video Gaming

There is a lot of precedent for what the metaverse will look like in the realm of video games.

Gaming in the future might look and feel different. Still, at its foundation, it's likely to be similar to how players currently immerse themselves in a game's world and atmosphere. So, what will be different? The main difference is in monetization.

So, what will be different? The NFTs, in the metaverse, will be applied to everything that can be tokenized, including in-game assets.

The gaming sector is a prime target for the "endowment effect" of NFTs.

Game creators and publishers have long leveraged gamers' desire to acquire extra powers, features, and assets as a means of selling their games. It's a historically effective strategy.

However, players may get frustrated when they realize their "purchases" only last as long as they keep playing the game they've "purchased" it in.

Your dazzling virtual objects disappear when you start a new game. NFTs are considered a solution to "address" the problem of players believing they "own" their in-game assets by making them sellable to others and transferable from one game to another.

Tokenized game asset exchanges are available. The misconception that a digital asset may be converted into a traded good using NFTs must be dispelled. Whether or not an in-game asset can be sold to another player depends on whether the game publisher supports the idea of tradability and has put in place the required infrastructure in-game to support it.

In theory, a games publisher that allows in-game asset trade may not need to tokenize its assets on the blockchain to do so.

Just because something is more complicated doesn't mean it's necessarily better. In addition, game publishers would be able to commission each sale and continue to monetize their assets, albeit from a new aspect, if they controlled their in-game marketplaces.

This in-game solution would significantly align with what happens when players "buy" and "sell" in-game assets.

In-game assets cannot be sold independently from their intellectual property, as stated in the NFTs section, and game publishers are not in the business of selling their intellectual property lightly.

The in-game assets also include licenses, not sales, as we said when talking about non-financial transactions in art. As explained in the game's terms and conditions, they grant you access to the object for a specific period and within a particular context.

Will in-game assets never be traded on NFT marketplaces because of this decision?

Probably not. NFT games are currently generating a lot of hype. Still, caution is advised because licenses are more difficult to sell than property rights.

This could lead to the demise of NFT games due to a lack of value.

A prime example of items that will eventually be traded in the Metaverse gaming NFT marketplace is portable game assets.

The ability to use your leveled-up rare sword, for example, from one game to the next, would be fantastic.

Can NFTs Make This a Reality?

Again, there's more to this story than meets the eye. Unlike in the real world, you can't easily pack up your blade and go on a trip with it. If the sword isn't in your host game, good luck using it to sever the heads of your foes. What's the point of creating a "foreign" word if the publisher already has some perfectly good ones available for you to use within their game world? You can't utilize a sword made for a video game in any other capacity. Nothing is more assured until the two corporations agree to make portability possible. There is little doubt that firms will take notice of the demand from gamers is strong enough. However, we believe that in-game assets (including characters) will only be transferred between games developed by the same company for the foreseeable future. It's important to remember that most gamers don't give a hoot about the legal ramifications of their purchases if they can have a fun time playing the game.

According to the endowment effect, there will undoubtedly be a widening gap between what games are composed of and what people believe they are.

This is a precondition for the metaverse to be a viable alternative to the real world; it should closely resemble it.

The metaverse gaming prospect relies heavily on advanced technology.

Fortunately, corporations do not have to spend as much money on infrastructure.

Processor and graphics technology advancements have been spurred on for years by a booming video game sector. Today, we are getting ever closer to photorealistic gaming experiences. It is just a matter of time before intellectual property and licensing issues take center stage in the gaming industry. There is no way around the inherent limits of that approach when applied to a notion of interoperability imposed by video games as a prototype for the metaverse.

Problems with NFT and related tokenization are more manageable than those connected to the metaverse's underlying infrastructure, in specific ways at least. Why should we assume that the metaverse would look like a single planet where everyone on it can connect and interact in all of these ways: love, hate, fight, reconcile, exploit, and heal?

Multiple metaverses, segregated at the very least by platform configurations but maybe also by content, genres, and publishing rights, are significantly more likely due to the intellectual property and the accompanying license.

The financial motive that has fueled technological growth is the construction of barriers between competing worlds. Metaverse's paradigm as a video game hints at the limitations built into the infrastructure that would create the virtual world. A metaverse that spans jurisdictions and platforms may exist. Still, it will be shattered by intellectual property laws, antitrust laws, privacy regulations, and the capitalistic ethos that has driven the video game industry for decades. When it comes to power, the metaverse's infrastructure will once again raise concerns about the amount of energy needed to run the CPUs and graphics chips.

Video games and the companies building the infrastructure that will support future generations of games and perhaps even a metaverse can serve as helpful guides. Sustainability and energy saving will be essential differentiators for organizations vying for market share in video games and platforms. People who want to make video games more immersive will need to be environmentally responsible (both in terms of energy usage and sustainable construction materials).

Game developers must consider green alternatives instead of simply creating more massive and voracious appetites for the earth's resources.

This is especially important since public opinion appears to be shifting toward a shared goal of preserving our planet.

Metaverse Gaming and Laws

Modesty has its bounds when it comes to the nature of human beings. Online video games and the platforms that host and market them teach us another important lesson: if left unchecked, they may degrade into hazardous environments.

Already countries around the globe are beginning to regulate the Metaverse gaming system.

For example, the EU Directive 2010/13/EU amendments seek to align nonlinear service regulation with linear TV restrictions to protect minors and harmful content and include specific video-sharing platforms (VSP) requirements to protect minors from harmful content.

Other European countries are also beginning to step up their regulation of the web.

Children (under 18) are the focus of the new ICO Age-Appropriate Design Code in the United Kingdom, which went into effect in September 2021.

The code recommends certain default settings for services that are likely to attract children, including considering children's best interests when designing any data processing in services.

A new German law, the Federal Protection of Young Persons Act (Jugendschutzgesetz - JuSchG), which took effect on May 1, 2021, aims to protect children and young people from harm caused by media consumption and ensure that media is only distributed or made available following the applicable age classification.

The various types of media and other publications that fall into this category include immoral and violent content; the detailed presentation of violent acts, murder, and massacre; and the recommendation of "the law of the jungle" to obtain 'justice.'

The French government has also enacted several laws that regulate online behavior.

One stands out: the pending French audiovisual reform draft law, which would combine the Conseil Supérieur de l'Audiovisuel (CSA) and the Haute Autorité pour la Diffusion des Oeuvres et la Protection des Droits sur Internet (HADOPI) into a single entity.

Among the many new powers that would be granted to this new "super-regulator," known as the Audiovisual and Digital Communication Regulatory Authority (ARCOM), would be the ability to regulate online platforms and combat harmful content on the Internet, and improve the fight against piracy.

It is uncertain if governments can successfully control and promote the moderation they now do in video games in the metaverse. Yet, it is possible in the real world.

If the idea of "platform" becomes nebulous, what liability might be imposed on a developer that does not implement anti-online harm moderation requirements on their platforms?

Would the regulators be required to interact with the public in the virtual world, like Agent Smith in The Matrix?

These issues will unfold in the coming years as the metaverse is continually being developed. All enthusiastic fans and loyalists of the concept can do is wait.

Chapter 4: Metaverse and Music

When compared to other industries, the music sector has always, historically speaking, been the first to react to any new internet invention.

Everyone is aware that the music industry was significantly disrupted and transformed beyond recognition in the early days of internet development.

As a result of the COVID-19 epidemic, the music industry, mainly performing artists, has been forced to innovate and find new ways to connect with their audiences.

As a result, they began to perform on the Internet.

To be fair, online streaming is not a new concept. Bands like the Rolling Stones were doing it as early as 1995. Streaming music is, in fact, the entire market model of companies like Spotify.

However, music consumption in the metaverse differs significantly from typical "vanilla" live streaming, or even subscription streaming, in several important ways.

Listed below are the differences.

The Ability to Create, Or to Perform In, A Virtual Venue

Using an avatar or other visual representation of the artist, sometimes mixed with an authentic video representation of the artist.

New production capabilities, such as manipulating the virtual environment and combining digital visual production with the artist's musical production

The Ability to Interact with The Audience In Real-time

The performance by Travis Scott on Fortnite was likely the most striking and commercially successful example of this revolutionary musical art form in recent years.

The performance by Travis Scott on Fortnite was likely the most striking and commercially successful example of this revolutionary musical art form in recent years.

This event generated a significant amount of attention and interest for this event.

Aside from virtual events and NFTs, another metaverse trend that has impacted the music industry is the emergence of virtual "artists."

The thought of listening to a virtual artist, who is made by artificial intelligence and does not have a real personality, may be repulsive to many serious music enthusiasts.

Despite this, there is no disputing that such musicians are gaining significant traction among young people who grew up with the Internet. The rapper FN Meka, who has been described as a "robot rapper known for his flamboyant style and Hypebeast aesthetics," is an excellent illustration.

While this may appear to be a frivolous, slightly futuristic bit of entertainment, it is built on a foundation of serious commercial possibilities. While writing this book, the virtual rapper has over 9 million followers on the TikTok video-sharing app.

Comparatively, Chance the Rapper, who is sometimes referred to as "one of the new crops of superstar rappers," had less than 2 million TikTok followers when writing this book.

These two incidents beg the question:

Is The Metaverse a Source of Opportunity Or A Source Of Danger For Music?

Both opportunities and threats for the music industry can arise from the metaverse, as demonstrated by the two cases presented above. Artistic careers are at risk if they rely on outdated methods that are no longer relevant in today's world of cutting-edge production and consumption methods and consumer experiences. For example, suppose you only possess the rights and monetize through subscription streaming channels. In that case, you won't be making enough money to justify your investment in these methods. They'll quickly become commoditized and automated.

Business opportunities are virtually limitless for those willing to push the boundaries and use all available technology to interact and create. Compared to online metaverse performances, even the most extensive arena tours cannot handle anything near the instantaneous, one-time global crowds the artist can attract to a live online metaverse performance.

The COVID-19 pandemic, which caused the entire world to shift to the Internet for entertainment, has demonstrated to the music industry that ticketed, well-produced, and compelling live streaming will be around for the foreseeable future. It is conceivable that the most significant concerts and festivals that take place in the actual world will in the future have an online component that is more committed, sleek, and transactional. Because of this, the metaverse will continue to exist in music for the foreseeable future.

Because of this, the metaverse will continue to exist in music for the foreseeable future.

What Are the Legal Ramifications of Music Being Played In The Metaverse?

When music is generated, played, streamed, and exploited online, rights clearances are the most important consideration, as they are in all aspects of the music industry. Most of the standard legal and licensing regulations for online exploitation apply in the metaverse, with some exceptions. However, music performance and exploitation in new, closed, or even open online environments add another potential layer of complexity to an already complex chain of rights in the music licensing process.

For example, a digital music service provider (such as Spotify) may promote and organize a live-streamed concert on a worldwide games console platform (such as the Sony PlayStation).

This concert could occur during the tournament's intermission being held and marketed by a leading games publisher (such as Electronic

Arts) who might be collaborating with a well-known brand during the interval of the event (e.g., Adidas).

Those interested in attending would need to be registered users of the gaming platform and have acquired entrance tickets to the eSports competition. Although the live-streamed concert would be available to a restricted number of superfans who joined a prize drawing by purchasing an original NFT token issued by the headline performing artist, the performance would only be open to the general public (for example, Drake).

Top-level prizes may include attendance at the live virtual event and an actual piece of digital goods.

Runners-up would still be able to watch the concert on-demand later, even though they'd be missing out on the thrill of a live show. The network of contractual responsibilities to negotiate and the rights-clearance concerns to consider, as illustrated by the example above, are not unlike the issues that lawyers may encounter in the real world when dealing with clients. The half-time show for the NFL Super Bowl is well-known in the music industry for being a highly prestigious but demanding production and clearance exercise that requires much planning and coordination. However, in many ways, the amount of complexity associated with clearing music for the metaverse can be substantially higher than the level of complexity associated with clearing music for the physical world.

Therefore, anyone wishing to use another's music in the metaverse must ensure that the terms under which they receive a license are compatible with where it is being utilized. While this appears to be straightforward in concept, a genuinely global virtual environment is governed in various ways depending on the legal jurisdiction. Censorship and content standards impacting a live performance by a Top 10 rap artist in the United States will be drastically different from those affecting a similar performance in, for example, Indonesia, Dubai, or Hong Kong. The political beliefs of artists are frequently expressed onstage.

These situations are more manageable in real life. Still, they are the stuff of nightmares for the legal compliance teams at large platforms, frequently tasked with maintaining positive relationships with local governments worldwide.

Who Is Responsible for Clearing The Permissions?

It may be claimed that customers are accustomed to the platforms themselves covering music licensing, at least when it comes to live performances or engagement with the public in the media. Twitch, Facebook, YouTube, TikTok, and PlayStation are online services that benefit from blanket agreements with music rights owners, collection organizations, and other online services.

At the very least, consumers can feel more confident about using music in the context they are operating, even though such sites' terms of service state unequivocally that music licensing falls solely upon the uploader. However, when music can be made, shared, and enjoyed in a real-time gaming metaverse or social setting, the situation becomes more complex and complicated.

By just establishing a meme, any user can now instantaneously control, tweak, and produce a whole new musical work that has the potential to go viral. These tools are publicly available and have the potential to cause widespread havoc.

The video-sharing app TikTok is unquestionably an essential platform for discovering and promoting new music at writing. Still, the users define whether a song will be successful more than ever. Because of the viral capacity of user-generated mashups and multiple synchronizations, lawyers who advise artists, labels, publishers, and even platforms themselves have an almost limitless number of opportunities for innovative licensing solutions, contentious disputes, and profitable transactional opportunities.

While the platform will be accountable for making reasonable attempts to get licenses for content posted by users, it will not be held liable for licensing copyrights in content brought to a platform by commercial operators (to put it bluntly). Suppose we apply this to the world of music. In that case, it instantly raises the question of whether a musician qualifies as a "professional user."

Artists as disparate as Ava Max, BTS, Marshmello, and Kaskade have performed through graphic representations in online gaming environments. At the same time, cutting-edge virtual reality services such as MelodyVR (now rebranded as the next generation "Napster") and Facebook's Oculus allow users to watch real-life concerts take place in a virtual reality format in real-time.

While there is no "one size fits all" method to securing rights for these types of events, there are several factors to consider:

- The person who is performing
- The legal framework under which the artist's recording and ancillary rights are controlled; the songs or works will be included in the performance.

It is essential to understand the following:

- Production components that are included (for example, choreography, which was previously the domain of only the most diligent of production rights clearance professionals, can now be a total minefield in the metaverse environment)

- The virtual engine that powers or underpins the production.

This includes the creative contributions of digital artists and other virtual participants.

Making Music in The Metaverse

Making new music in the metaverse will be a rewarding experience.

It goes without saying that if people begin to reside in the metaverse, project their image, and spend their time there, the next logical step for them is to transition from the real-world recording studio to the virtual creative environment. There are numerous examples of this already taking place in the world. A wide variety of virtual reality headsets and controllers that allow users to interact with graphical interfaces that simulate musical instruments are currently available. The air guitar transforms into a real guitar - Rock Band VR is on the horizon.

In this digital age, it is now feasible to form your band online and convert yourself from a balding, middle-aged "dad bod" into a lavishly coiffured, tanned, lithe rock hero who lives out your fantasy of playing guitar in front of large crowds. On a more practical level, metaverse environments such as Minecraft, Roblox, and Fortnite incorporate song codes, instruments, and recording facilities and controls for manipulating music, allowing players to express themselves musically. While the vast majority of this activity will result in original copyright that has little or no monetary worth, users have countless opportunities to infringe or encroach on well-known unknowingly, commercial music or assets, which could result in legal action.

Do you want to listen to some Frank Sinatra crooners in an electric jazz modern remix with your virtual buddies in the metaverse?

It's not a problem.

Of course, as the mix of innovative technology, people, and connection progresses, the complexity of the legal challenges also increases. Music is already one of the most convoluted, complex, and divergent aspects of entertainment law, and this is only the beginning. The prevalence and expansion of music in the metaverse indeed present new challenges. Still, it also offers enormous opportunities for lawyers to innovate and assist their clients – not only in navigating through the existing frameworks but also in developing new models and methods of exploitation of copyrights that contribute to the creation of incremental revenues and value for the industry, as well as for the platforms that invest in the metaverse itself.

Chapter 5: Metaverse and Artificial Intelligence

In recent years, artificial intelligence (AI) programs have gained the ability to behave intelligently and make music, art, and other forms of original creative output.

For $432,500, Christie's auctioned a portrait of Edmond de Belamy made by an artificial intelligence system three years ago.

SONY CSL Research Lab has also developed an artificial intelligence system called Flow Machines.

This AI program can compose new music based on anything from the Beatles to Bach, among other things. The Metaverse, whether it's an extension of the real world or any number of computer-generated worlds, is bound to include an overlying layer of unfathomably vast amounts of "data."

As a feature of that data, a person or entity creating and controlling an environment known as "the metaverse" will generate and disseminate that data.

Nevertheless, unlike the physical world, the Metaverse is wholly artificial. If a digital tree or cloud does not "belong" to its creator in the Metaverse, it will not exist in the Metaverse. We may anticipate that practically everything in the Metaverse, from the appearance of our avatars to the clothing we wear and the vehicles we drive, will be the intellectual property of someone. Artificial intelligence (AI) uses machine learning technologies to study, digest, and analyze massive amounts of data to develop rules of application, which are referred to as algorithms. The examination of fresh data sources and the observation of its own data output allow machine learning software to improve itself after it has been "trained continuously." A new branch of artificial intelligence has emerged in recent years, encompassing computing systems that try to emulate the function of the human brain in evaluating and processing information. These systems are referred to as artificial neural networks. They also include coupling computer networks in generative adversarial networks, in which the computers learn from one another. Several debates have erupted about AI machines' tremendous data consumption and their art in recent years. Is it possible for artificial intelligence to digest vast databases that contain copyrighted works and then use machine learning to "create" original works without infringing on the rights of third parties? Are the results created by AI protected under intellectual property laws? Machine learning and fair usage are two important concepts to understand. In their endless search for, digestion of, and aggregation of content, AI search engines inadvertently consume copyrighted items such as music videos, songs, novels, and news stories as they crawl around the world wide web. The legality of this digesting, which is usually conducted without the copyright holder's authorization, depends on whether it falls inside an authorized exception to or beyond the scope of copyright law.

The "fair use" exception to copyright law in the United States is the most invoked. As defined by Section 107 of the Copyright Act, "fair use" is determined by considering the following four factors:

1. The purpose and character of the use.
2. The nature of the copyrighted work.
3. The amount and substantiality of the portion used in relation to the whole.
4. The effect of the use on the potential market for, or value of, the copyrighted work.

It is expressly permitted by Section 107 to make fair use of a copyrighted work for teaching, scholarship, or research purposes.

Fair use is determined by many factors: whether the use is "transformative," as determined by the courts. A fiercely discussed topic that will have ramifications for the future of intellectual property law is whether machine learning of copyrighted information qualifies "fair use."

The future of Artificial intelligence in the Metaverse depends on the interpretation of copyright laws.

A good scenario is Thomson Reuters and West Publishing Corp. vs. Ross Intelligence, Inc. The law firm sued the IT company, alleging that it used machine learning to construct a legal research platform for Ross using the Westlaw database.

Will this be allowed? Will fair use protect machine learning?

A court recently determined that Google Books' scanning of more than 20 million books, many of which were subject to copyright, constituted a "non-expressive" and transformative fair use of the texts. The reasoning behind that decision was that it enabled users to find information about copyrighted books rather than the expressions contained within the books themselves.

Protection may be available if the use of copyrighted content is "non-expressive" fair use, as opposed to "expressive."

Mechanical digestion of copyrighted materials may be permissible if the artificial intelligence (AI) utilized in machine learning is not "too sophisticated." Of course, artificial intelligence has progressed well beyond Google Books. AI can now learn how authors communicate their ideas and then generate their original creative output. This expressive machine learning may, in turn, harm the market for works written by humans.

This expressive machine learning may, in turn, harm the market for works written by humans. Because AI can produce outputs like human expression and personalization, the use of copyrighted works for machine learning may result in copyright infringement, especially if permission has not been secured from the owners of those works before the use of those works.

Metaverse content is being used to train artificial intelligence. This "intellectual property everywhere" scenario is likely to impact how we access and re-use the data created within the Metaverse in the future. As examples of technology whose ability to operate may be hampered in an "intellectual property everywhere" scenario, artificial intelligence (AI) and machine learning (ML) are excellent examples of technology whose ability to operate – given their reliance on ingesting vast amounts of data – may be hindered in an "intellectual property everywhere" scenario.

Data and information used to train a machine learning model may be subject to restrictions in the future. Not all information is "protected" or "owned" - for example, protection is unlikely to extend to historical meteorological data, pollution levels, the structure of clouds, or the sound of birdsongs, among other things. Every bird song in the Metaverse is likely to be the work of a computer that a person created, and as a result, it may be able to be protected (for instance, the code used to write the song may be protected, or a human writes the song itself). This could lead to the emergence of new and exciting legal conflicts.

For example, in a world where "intellectual property is everywhere," using nearly any type of information in a machine learning system would almost certainly be considered restricted conduct for which authorization would be necessary.

For example, just "reading" material should not be regarded as a restricted act when it comes to copyright. Still, acts of copying or reproducing – which are likely to occur in the real-world functioning of a machine learning system – almost certainly are, unless a relevant copyright exception is proved to apply, such as the doctrine of fair use in the United States, notable machine learning exceptions in jurisdictions such as Japan, or the more limited (and highly competitive) concept of fair dealing in the United Kingdom Another certainty of the Metaverse is raised by the final point made.

Applicability of fragmented and diversified national intellectual property systems to "international" machine learning and output distribution will be at least as difficult as it has already proven to be in the context of traditional content distribution over the Internet. This pattern of territorial arbitrage that has marked the evolution of the Internet will undoubtedly reappear in the Metaverse; it is almost likely.

Is AI-created output infringing? Even if the creation of the AI machine learning model in and of itself is not infringing, if the output generated by an AI system that has been trained on a particular type of data is substantially similar, it may be an unauthorized "derivative work" that infringes copyright in the preexisting works. For example, companies like Jukedeck, which was purchased by ByteDance and taken off the market, have used machine learning on recorded music to create algorithms that, in turn, create new music. Because of the potential for companies like Jukedeck to generate automated music that would hurt the market for music composed by humans (such as production music typically used in film or television), these creative outputs will almost certainly receive heightened scrutiny.

Do intellectual property rights protect AI-generated content? In the Metaverse, artificial intelligence (AI) inventions will almost certainly make up a significant portion of the environment – sometimes literally, as in the instance of the Azure-driven location models and maps generated by Microsoft Flight Simulator.

The ownership and rights issues in the outputs of artificial intelligence systems present their own set of issues. Copyright in creative work (and, as a result, its "ownership" and protection) are preconditions for the existence of copyright in creative work (and, as a result, for its protection and "ownership," according to international law. These principles fall apart when the link between a human author and the creative work is broken — most notoriously in the "monkey selfie" case, where an image shot by a monkey was deemed not protected by copyright.

Artificial intelligence-generated outputs (depending on the circumstances, can be distinguished from works made by AI aid) call into question norms that solely consider human authorship of copyright works. Even the United Kingdom's one-of-a-kind provision governing "computer-generated works," under which the person "by whom the arrangements necessary for the creation of the work are undertaken" is deemed the author, confirms the importance of identifying a human rather than a computer as the author of a "creation".

Additionally, traditional reasons for copyright protection, such as rewarding the creation of works or preserving the natural rights of authors, are rendered ineffective when the creator is a computer that requires no incentive and does not have a distinct personality. In short, the legal system in the United Kingdom does not appear to be welcoming or accommodating of robot-generated inventions, which (at least for the time being) look destined to fall into the category of free and free-flowing information.

Do you think an AI-generated metaverse has the potential to reshape our world by creating a wonderful environment for the public domain and "commons" to thrive? The question is whether or not an AI-generated metaverse can compete with human-generated worlds in a massive conflict of intellectual property fights. It's possible that the android's doodle of an electric sheep was created by someone else and is not protected by copyright, but the android's programmer may still wish to license it to you.

In the United States, the fundamental goal of copyright legislation is to encourage the creation of new works of art by providing writers with a financial incentive to protect their creations under the law. This economic incentive is provided to authors for the benefit of the public since enabling authors to be compensated economically for their works would produce more innovative content on the Internet.

Will artificial intelligence firms be able to benefit from the economic protections afforded by copyright if they continue to invest in the technologies required for the machine-based production of creative works?

According to Section 102 of the Copyright Act, a work must be "an original work of authorship fixed in any tangible medium of expression now known or later discovered..." to be copyrightable.

The need for human authorship is not explicitly stated in either the Copyright Act or the United States Constitution. Still, the courts and the Copyright Office have worked on this assumption. Copyright Office practices require human authorship to register works created entirely by mechanical methods. The Compendium of Copyright Office Practices includes a requirement for human authorship to register works. This case was brought against a wildlife book publisher and dismissed by the Ninth Circuit three years ago because an author who was not human did not have the standing to sue under the Copyright Act. The selfies were taken by a crested macaque monkey and published in the book by a wildlife photographer. This means that once developed, AI-generated works will become part of the public domain and will be available for free distribution.

As things stand, this has significant ramifications for the creation of artificial intelligence-generated works because the firms and investors who fund the machines that make them are currently not protected by copyright laws in the United States. There has been a great deal of debate about whether copyright laws in the United States will evolve to provide this level of protection.

It has been argued that other non-natural persons have been granted legal rights, which supports the extension of copyright protection to nonhuman authors. For many years, corporations in the United States have enjoyed the same rights to contract as individuals and the ability to enforce contracts to the same extent as individuals, in addition to the need to pay taxes. The concept of machine-based work-for-hire doctrine has been advanced by commentators, who argue that the end-user of an artificial intelligence program that generates creative content should be considered the owner of that content.

According to these commentators, the AI program is viewed as the equivalent of a contractor hired by an employer to produce content that the employer owns. However, some have argued that the end user's creative inputs justify the end user's status as a creator of AI-produced content. In contrast, others argue that the AI program should be viewed as a tool for the end-user. 22 Artificial intelligences as a copyright enforcement tool Machine learning, in addition to providing human authors with the power to create new works of art, also offers them the ability to enforce their rights and monetize their works of art better.

Many companies, like Audible Magic and Google, have created artificial intelligence software that detects material and assists in the detection of suspected copyright violations. It is expected that these technologies would provide significant economic benefits to human authors. Should artificial intelligence copyright be based on originality? Some countries, such as the United Kingdom, have taken steps to protect computer-generated works based on the components of creativity embedded within the work to stimulate investment in artificial intelligence (AI) technologies. Certainly, as artificial intelligence advances and generates more "creative" works, the discussion over the ability to copyright these works and who has ownership rights will continue to rage.

Other topics that are receiving a lot of attention in machine learning and artificial intelligence are the ethical compliance of AI systems, as seen by the increasing number of publications and debates in this area. The moral repercussions and hazards of artificial intelligence (AI) are currently believed to be very application-specific. As an example, the potential for in-built biases of an artificial intelligence system to have severe effects on human subjects is thought to be far more apparent in the context of criminal justice applications than in the context of an artificial intelligence-generated piece of artwork. This is at the heart of the European Commission's latest draft Artificial Intelligence Regulation, which identifies "high risk" AI applications that must be subjected to regulatory criteria. Suppose we are to ensure that a Metaverse is safe for everybody. In that case, every AI-generated three-dimensional game environment will likely be free of biases, bullying, and other artificial expressions of violence, which are all too often in our real-world environment in the future. If that day comes, all artificial intelligence operators will likely be required to consider their internal processes and governance in light of the high level of safety and security required to enter the Metaverse's construction site. When it comes to ensuring that humans feel comfortable, safe, and at ease in the Metaverse, certain factors should be considered.

Considerations such as the potential for bias in systems and outputs, the quality and nature of training data, the resilience and accuracy of systems, and human oversight and intervention are all essential considerations to bear in mind.

The European Union's Attitude to Artificial Intelligence And The Metaverse

There is no formal EU legal framework for regulating artificial intelligence and the Metaverse. Artificial intelligence development, implementation, and usage are governed by various horizontal laws and principles, including data protection and privacy, consumer protection, product safety, and legal responsibility. The European Commission, on April 21, 2021, announced their long-awaited proposal for a law on artificial intelligence, with the goal of making Europe the global center for trustworthy artificial intelligence (Proposal for a Regulation laying down harmonized norms on artificial intelligence (Artificial Intelligence Act)). The proposal represents the culmination of several years of preparatory work by the European Commission, including publishing a "White Paper on Artificial Intelligence" in 2012. According to the Commission's vision, the basic rights of individuals and enterprises should be protected and strengthened while artificial intelligence (AI) innovation is encouraged throughout the EU.

Who Is It That the Proposition Is Intended For?

AI providers and users in the EU and providers and users in a third country where the system's output is utilized in the EU would be subject to the new proposed legislation, regardless of whether those providers are situated in the EU or a third nation. What exactly is contained within this proposal? The Commission takes a risk-based but overall cautious approach when it comes to artificial intelligence. While it recognizes the potential of artificial intelligence and the numerous benefits it offers, it is also acutely aware of the dangers these new technologies pose to European values as well as fundamental rights and principles. They adhere to a risk-based approach that may be broken down into four main categories:

1. Unacceptable risk: Artificial intelligence systems that are deemed to pose an obvious threat to people's safety, livelihood, or rights are generally prevented from being developed. The possibility of psychological or physical injury arises, especially when systems or programs manipulate human behavior to impact the user's free will, resulting in an intolerable danger. For example, toys that use voice assistance to urge youngsters to engage in potentially risky activities would fall under this category of products.

2. High risk: Artificial intelligence systems that have been recognized as high risk are permitted, but they are subject to additional regulations and conformity tests. Among these systems are artificial intelligence (AI) technologies, which are used in a variety of fields that require higher levels of protection, including education, critical infrastructure, employment management, product security components, law enforcement in cases of interference with people's fundamental rights, and asylum and border control management.

Here are only a few examples of unique responsibilities:

1. Before being placed on the market, the systems must undergo a thorough risk assessment and mitigation process.

High-quality data sets, complete documentation on all information essential on the system, and its intended purpose must also be provided so that authorities can assess compliance with the requirements. The systems must match the user's needs in terms of transparency and information, and humans must oversee them reduce the chance of failure. This category includes all remote biometric identification systems, subject to the same stringent regulations as the rest of the industry. It is generally unlawful for law enforcement officers to use them in real-time in publicly accessible areas for law enforcement purposes.

2. Only a small number of rigorous exceptions are permitted, and a legal authority must approve these

3. Artificial intelligence systems with modest hazards usually are approved, but they must also comply with stringent disclosure requirements.

Users of artificial intelligence systems, such as chatbots, should be made aware that they are talking with a machine to make an informed decision about whether to continue or cease interacting with the system.

4. Minimal risk

The great majority of artificial intelligence systems, such as video games or spam filters, fall into this category and are legally permitted to operate since they pose minimal or no harm to the rights or safety of users.

What Comes Next?

The European Commission's 108-page proposal attempts to govern a new technology before being widely used. As the world's most aggressive watchdog of the technology industry, the European Union may serve as a model for comparable measures in other parts of the world. There are significant ramifications for major technological businesses that have invested substantial sums in artificial intelligence development and many other organizations that use the software to produce medication or assess creditworthiness. Versions of the technology have been utilized by governments in criminal justice and the distribution of public services such as income support. With such a broad definition of artificial intelligence systems, the rule is sure to have a considerable impact across all industry sectors, particularly in those industries that wish to be successful in the Metaverse. After that, the proposal will be forwarded to the European Parliament and the Member States for consideration under the standard legislative procedure. Given the contentious nature of artificial intelligence (AI) and the enormous number of players and interests involved, it appears unlikely that this will be a simple or straightforward procedure. There will almost certainly be numerous modifications, as well as, hopefully, some more clarifications. It is intended that, once the law is adopted and passed, the rule will be directly applicable in all of the EU's member states.

Chapter 6: How to Invest In The Metaverse

Like any other investment opportunity, Investing in the Metaverse requires careful calm consideration and logic.

Investors are encouraged to think carefully and deeply before deciding on which investment strategy to employ. This chapter will talk about the top 5 investment strategies you can use to invest in the Metaverse wisely.

Keep reading.

What Are Investment Strategies?

Investment plans are techniques that assist investors in determining where and how to invest their money based on their projected return, risk appetite, corpus amount, long-term versus short-term holdings, retirement age, the industry of choice, and other considerations. Investors can tailor their plans to meet their specific aims and aspirations when it comes to investing.

Investment strategies can be divided into five types. Let's take it one at a time and go over the many types of investment methods.

Passive and Active Investment Strategies

The passive technique is purchasing and storing coins rather than exchanging them regularly to avoid greater transaction costs. Because they believe they will not outperform the market due to its volatility, they prefer passive tactics, which are less hazardous. On the other hand, active tactics entail frequent purchases and sales of products. Investors feel they can outperform the market and earn higher returns than ordinary investors believe.

Growth Investing (Short-Term and Long-Term Investments)

Investors choose the holding term based on the amount of value they want to add to their portfolio. To increase the corpus value of a token, investors must believe that the token will grow in the upcoming years and that the intrinsic value will increase. This is referred to as growth investing in some circles. On the other hand, short-term investment is made when investors feel that a token will offer good value within a year or two of the investment. The preferences of the investors themselves also determine the holding time. For example, how quickly they require money to purchase a home, send their children to school, or fund their retirement plans, among other things.

Value Investing

The value investing technique includes investing in a token based on intrinsic value rather than the market value since markets undervalue such companies.

Investors in such companies hope that, when the market undergoes a correction, the value of such undervalued companies will be corrected, causing their prices to skyrocket, providing them with substantial profits when they sell their shares. Warren Buffet, the world-famous investor, employs this method.

Income Investing

Coins in this sort of approach are chosen for their ability to provide cash flow rather than for their ability to grow the overall worth of your portfolio. Cash income from this form of investing can come in two types: fixed income (through yield farming) or dividend income (staking)

This technique is preferred by investors searching for a consistent stream of income from their investments.

Investing in Dividend Growth Companies

Tokens with a track record of routinely paying interests are more stable and less volatile than other companies, and they strive to improve their dividend payout each year. In this strategy, the investors reinvest the earnings and reap the benefits of compounding over the long run.

Investing Guidelines for Beginners

The following are a few investing tips for beginners that you should consider before making a financial investment.

Set Financial Objectives

Establish financial objectives for how much money you will require in the upcoming period. This will enable you to determine if you need to invest in long-term or short-term investments, as well as the amount of return you may expect to receive.

This will enable you to determine if you need to invest in long-term or short-term investments, as well as the amount of return you may expect to receive.

Investigation And Trend Analysis

Before investing, take the time to thoroughly research how the market operates and how various financial instruments function.

Additionally, analyze and monitor the price and return trends of the coins you intend to invest on.

Portfolio Optimization

Choose the most appropriate portfolio from the portfolios that best fit your objectives. An ideal portfolio generates the most significant return while posing the least amount of risk.

Risk Tolerance

Determine the level of risk you are willing to accept to achieve the desired return. This is dependent on your short- and long-term objectives as well. You would seek a more significant rate of return in a shorter period, while you would seek a higher risk in reverse.

Risk diversification is essential. Invest in different projects to diversify your risk and increase your returns. In addition, be sure that both tokens are not associated with one another.

Advantages Of Investment Strategies

Investment Strategies Have several Advantages. Some of the advantages of investment techniques include the following:

- The use of investment methods, which invest in various investments and industries based on time and expected returns, allows for risk diversification in the portfolio.

- When constructing a portfolio, investors can choose from one strategy or a combination of methods to meet their specific tastes and requirements.
- Investing wisely allows investors to make the most of their money and maximize their returns.
- Investment techniques can help you save money by lowering your transaction costs and paying less tax.

Limitations To Investment Strategies

Examples of investing methods' drawbacks include the ones listed below:

- Investors of average means have a difficult time outperforming the market. Even though cryptocurrency is a great project to invest in with remarkable and fantastic rewards, many new beginners still struggle to see tangible results in the sector due to a lack of patience and greed.
- Even though a great deal of study, analysis, and historical data are considered before investing, most decisions are predictive.
- It is possible that the results and returns will not be as expected, which will cause the investors to be further behind in accomplishing their objectives.

- It is critical to have a well-thought-out investment strategy. It will assist you in weeding out bad portfolios and boost your chances of becoming successful.
- Consider a few fundamental questions, such as how much money you want to put into it. What kind of return do I require? What is the extent of my risk tolerance? What is the length of my investing horizon? Why was it necessary for me to invest?
- The more specific your objectives are, the more confident you will be in your ability to make sound investing decisions. Always be on the lookout for potential investment opportunities and never make a large sum of money at once.
- Building a portfolio is like building a house from the ground up, brick by brick and dollar by dollar. It is best to do it right rather than do it fast.

Chapter 7: The Metaverse and Virtual Estate

The Metaverse is a virtual land that exists only in the user's mind.

The Metaverse is ready to cause significant disruption in even the most unlikely industries, such as the real estate industry. Through its lack of connection to the physical world, the metaverse is poised to displace some long-held legal notions, such as the concept of property ownership, from our human society.

Real-world geolocations can be linked to user-created digital settings in the Metaverse, a virtual realm that lies between the real world and virtual reality. These surroundings will be available for purchase, sale, and ownership soon and extensive customization.

The decision to purchase Metaverse land is big for many people, but the benefits outweigh the risks.

This chapter will cover the process of acquiring land in the Metaverse.

Land In the Metaverse

Is there anything special about Virtual Land, and what can you do with it?

True, it might "just" be a virtual metropolis, but investors are forking over real money to buy land in it. Buyers in Decentraland have the freedom to build anything they wish on their parcels of land. To many, trading products and services for bitcoin is a way to generate money in the virtual world. The American dream is taking on a new meaning in virtual worlds, where investors purchase pieces of property and strong communities increase desirability.

The American dream is taking on a new meaning in virtual worlds, where investors purchase pieces of property and strong communities increase desirability.

Why do People buy Virtual Land? People buy virtual land for a variety of reasons.

The following are a few of the reasons why purchasing virtual land is deemed necessary:

New Asset Class

Digital real estate has proven to be a valid asset class. Its value is expanding exponentially, which makes it an appealing investment possibility. There is also a possibility that someone may turn it into a viable financial asset, analogous to real-world art and actual-world real estate.

Fear of Missing Out (FOMO) (Fear of Missing Out)

People who purchase wonderful reals or virtual do so because they have a horrible sense of missing out on something amazing. So many individuals miss out on buying Bitcoin when it was so cheap has prompted them to explore alternative items such as virtual real estate.

Exceptional Profits

Because of its association with the constantly increasing crypto-investment world, virtual land has the potential to generate tremendous returns. Many people have been able to gain thousands of dollars in a short amount of time because of the ease with which they can flip land (like they do with real estate) and the consistency of the bull market.

Possibilities For Additional Earnings

In the long run, virtual land opens new possibilities for what can be done with the property, such as constructing art galleries, executing advertising campaigns, or simply renting the land to others to earn money from their construction projects.

Some users utilize their virtual property to build virtual casinos, which they can play in.

In addition, major retailers are looking into the prospect of opening storefronts in virtual reality (VR).

It Is Less Complicated Than Purchasing Real Estate

There are also significant advantages to using digital real estate over conventional methods, including eliminating time-consuming paperwork, property maintenance, and tax payments. Furthermore, the usage of blockchain technology enhances the security and traceability of land acquisition transactions.

Low Entry Barrier

Prices for real estate are rising worldwide, yet virtual properties provide similar benefits for less than one percent of the cost. Due to the high cost of acquiring natural land, this goal may be out of reach for many people, but purchasing virtual land is lower.

What Is the Best Place To Purchase Virtual Land?

Whether we're strolling down the street or looking out the window, we're surrounded by virtual worlds that are constantly changing and evolving. More and more companies are stepping forward to construct virtual worlds that can be viewed on compatible devices like PCs, mobile phones, or even appropriate headsets since the emergence of Virtual Reality and Augmented Reality technologies.

Buying Virtual Land in The Metaverse: A Word Of Caution

To a large extent, earnings outweigh all other considerations when it comes to purchasing virtual estate (as well as a cryptocurrency or NFT).

Numerous newcomers to this ecosystem are looking for other means of earning money in the wake of the coronavirus outbreak. Even though some people are simply trying to save up considerable sums of money obtained via cryptocurrencies like Bitcoin, Ethereum, and Solana, these factors certainly contribute to the current bull run for virtual land.

Although it may appear to be a pointless endeavor to some, people are spending millions of hours every day in virtual worlds despite this.

Thus, it will retain its value as long as others believe it is valuable. Some short-term investors will indeed be interested in profiting off virtual plots in the hopes that their value will increase in the future. Another group of people, on the other hand, will just seek access to these plots because they love doing so.

Chapter 8: Top 5 Crypto Metaverse Projects

This chapter will discuss the top 5 cryptocurrency projects relating to the metaverse and how to invest in them.

1. SANDBOX

What Is the Sandbox token and How Does It Work? The Sandbox (SAND) is an Ethereum-based ERC-20 token that serves as the native asset of the Sandbox virtual economy. It is traded on the Ethereum exchange.

SAND tokens will be 3 billion, with around 900 million tokens already in circulation.

For a project that prides itself on being decentralized, it is surprising to find that most of the entire supply is allocated to the corporation, its personnel, advisors, and investors. In addition, the token performs two essential services on the platform at the time of this writing. In-game things can be purchased in the marketplace with this currency, staked to generate interest.

The Sandbox team's editing program is the backbone of the game. The software allows users to construct 3D objects such as creatures, costumes, buildings, automobiles, and pretty much anything else they can think of in their minds. A non-fungible token (NFT) is created for each item, which may then be sold on The Sandbox's marketplace or secondary markets such as OpenSea.

The ability to develop whole 3D games that can be played within the virtual world without knowing how to code is also available. You can monetize your games and earn SAND every time someone plays them if you're determined to make money from this enterprise.

The Sandbox game is played on a vast map divided into segments, which the developers refer to as LAND.

A single NFT is worth thousands of dollars, and each piece is sold as a standalone. Many gamers have turned into virtual real-estate tycoons, enhancing their plots with stuff purchased from the marketplace and reselling them.

Lands can be combined to form larger and more valuable estates and districts due to their geographic proximity. It is possible to make money with your land in ways other than simply flipping it, so explore your options. You can host events and games, charging a price for admission and hoping to attract many paying customers.

According to the inventors, SAND will eventually serve as a governance token and a cryptocurrency.

According to the project's whitepaper, a decentralized autonomous organization (DAO) will be integrated into The Sandbox by 2023, at the earliest. The SAND holders would vote on issues that affect the future of the game and ecosystem once this process is completed.

It's difficult to conceive an utterly decentralized Sandbox because the bulk of tokens and voting rights are likely to remain in the hands of the company, its crew, and its backers. A significant portion of the platform's token supply may be made available to the public in the future. However, as of now, the token's decentralization is still in its infancy.

The Price History of The Sandbox

The announcement by Facebook to rebrand itself as Meta was by far the most significant event in the history of The Sandbox's price.

The token had only achieved an all-time high of $1 in September, just before the announcement. SAND had been trading at around $0.05

per share for several months until late January 2021, when it began to join the rest of the market in the incredible 2020-2021 bull-run that started in late January. It reached its first significant high in March, at approximately $0.85, just a few months before Ethereum reached its all-time high in May.

Even though its rise to $0.85 and then $1 was spectacular, it was outdone during the first week of November, when it soared to $3 in a matter of days.

Following Facebook's announcement, the value of tokens such as SAND surged. On the surface, this attitude appears to be completely logical.

As more people become familiar with the technology, the existing platforms may see a significant increase in use. In contrast, Facebook has now announced that it will be developing what will undoubtedly be the most formidable competitor to The Sandbox. A tech behemoth has entered the space, most likely intending to destroy the existing ventures with a figurative steamroller.

Decentraland (which also features later in this chapter) is expected to be Meta's most significant competitor until Meta's Metaverse is released.

On the surface, the two initiatives appear to be highly similar.

These two games are metaverse games centered around a cryptocurrency and have an NFT-based economy in common. However, there are significant distinctions between the two of them. Using their groundbreaking 3D editing software, The Sandbox allows its customers to create practically anything they can imagine.

Decentraland is a little more straightforward for the typical person, but that may be more appealing to them.

Decentraland also has a significantly higher number of active users at the time of writing and appears to be receiving a little more brand recognition. For example, Coca-Cola Co. (NYSE: KO) has decided to begin selling branded NFTs in the game as early as this year. Although the Sandbox boasts many noteworthy partners, such as Atari and Snoop Dogg, it appears to be falling behind its competitor in this area.

How To Buy Sandbox

Because the Sandbox token is a relatively popular cryptocurrency, it may be found on several notable cryptocurrency exchanges. FTX and Gemini are two of the best trading platforms for the token, and they both support it. It is also sold on other exchanges like Binance and KuCoin.

The success of SAND will most likely be determined by how well players receive it in the future. Even if the Metaverse becomes a hugely popular concept, The Sandbox will require people to participate in the game and contribute to its ecology to succeed.

The Sandbox has the potential to be a terrific investment. For confident investors, it has already been shown to be such.

If you had purchased it in January, your investment would have increased by about 5,000 percent. It is, nevertheless, an exceedingly risky investment, as are most cryptocurrencies.

If the game cannot attract users and investors, the price of SAND will almost certainly fall over time. Whether The Sandbox will be a major player in metaverse games within a year is tough to predict. Still, it is likely a better bet than some of its lesser competitors.

In What Way Does Purchasing a Piece Of Land In A Sandbox Benefit You?

A decentralized community-driven gaming, visual art, and game design environment built on the Ethereum blockchain, Sandbox allows makers and designers to create and commercialize their new functionalities (NFTs), art experiences (including gaming observations), and gaming observations (including NFTs).

One of the most important goals of Land areas is to provide a platform for game developers and designers to expose experiences on them that can be played and monetized by gamers.

Various other services, such as leasing out land and making land claims, will be available.

There are two types of land available for purchase: standard and premium.

How To Buy Land on Sandbox

Sandbox sells land via public auctions of real estate. The announcement is made in the official communities in front of the public.

Before purchasing a plot of land from the Sandbox, you must first register with the site.

The map may be found on Sandbox's official website, where those interested in purchasing land during public property sales should go.

To purchase a piece of land, select it from the list of available parcels.

It will be indicated in yellow if there is any accessible Premium land. In contrast, it will be displayed in grey if there is any available standard land.

The "Buy" button, highlighted in blue, should be used if you wish to acquire the land.

Until the sale is finalized, canceled, or fails to proceed (for example, because of a lack of natural gas), the land would be held in reserve (after two hours).

After then, the land would turn purple, signifying that it had been reserved.

Your bank account should indeed be displayed, prompting you to complete the payment and explicitly explain the quantity of gas (charged in ETH) that you are now spending in your transaction.

It will be completed as soon as your bank account receives confirmation of the payment. The amount of gas you choose and any blockchain congestion significantly impact the time it takes to execute the transaction.

When you are successful, the land will become red to indicate that you have now acquired ownership of it.

How Does Buying Land on Sandbox Make You Rich?

Lands on the Sandbox, a type of digital real estate that allows you to generate a very clean and consistent source of revenue, will give you multiple options to make a very clean and constant stream of income.

Let's look at how the land can make you a very wealthy individual.

Hospitality Industry

Users can host living experiences like games, art museums, stores, scenery, engaging education, and other activities on Sandbox. The primary role of property is to allow users to host live experiences.

Sandbox's own game development program, known as the Game Maker, may be used to design and build these experiences, which can then be made available on any of the creator's territories. Players may be required to pay an entrance fee in cryptocurrency to access the experience offered on the land.

Staking

Landowners will be able to stake cryptocurrencies on their lands in exchange for passive incentives in the future, thanks to a feature planned for Sandbox.

One of these perks is GEMs, an ERC-20 token that is extremely valuable and sought after by asset design professionals.

As a bonus to the regular staking incentives, these GEMs can be sold on the open market in exchange for cash. Your Land area ownership functions as a multiplier when you deliver SAND-ETH cash flow to a UniSwap liquidity provider, increasing the amount of SAND cryptocurrency you receive from cash flow mining.

It will also be possible for renting landowners to lease their properties to third parties, such as game designers and film production firms, who may have missed out on a property during the original sales.

Upon the completion of the sale of all available land, the quantity of available land for rent will surge as more individuals become aware of The Sandbox and opt to submit an experience there.

Contests and Giveaways

Organizing contests and giveaways on land might bring a significant number of paying buyers to your property to participate in the competition or giveaway. Other people may conduct tournaments or give away prizes on your Lands to promote their businesses and increase your exposure.

Advertisements

Are you an affiliate marketer? Do you own a business of your own? Are you a published author? Are you a creative person? For that matter, anyone interested in selling a product or service. Then why not widen your reach by utilizing some of your advertising space on your property to market yourself to players and visitors who might otherwise be unaware of your product or service.

Asset Non-Financial Transactions

Whenever you decide to publish an experience on your property, you will develop admittance requirements and a price for visitors to participate in it.

It is possible to impose an entrance requirement that players first hold a specific asset, such as NFT, as one of the prerequisites. Suppose a player wishes to participate in your land-based swashbuckling pirate game but hopes to purchase a weapon from your worldwide NFT sword collection, which you also released on the global market.

Land For Sale in The U.S.

The sale of land inside Sandbox is, of course, an additional opportunity to generate money, mainly if the land is in a high-traffic and highly sought-after zone of the Metaverse.

In the long run, however, if you can maintain your composure and hold onto your Land regions for a longer period, you will most likely earn more money through a combination of the various strategies discussed in this article.

What Is the Cost Of Land On Sandbox In Dollars?

The cheapest land available for purchase on Sandbox is more than 10000 USD.

The average price of land has increased significantly in the last three months. The average land price increased from 40 USD to 960 USD in less than a year.

Is There Risk Involved?

There is, without a doubt, one. As a result, proceed with caution. Land prices may decline in the future. It is still in its early stages and has not yet gained widespread acceptance. However, given the current trends, it is doubtful that we would see a significant decline in the price of land. However, now would be an excellent moment to purchase it if there is. Sandbox is a company in which I have a lot of confidence. Looking forward, it will be intriguing to see where life takes us.

The Metaverse is the future, and any investment in land within Sandbox or other Metaverses could turn out to be a terrific venture soon. The reward to risk ratio is high. According to estimates, the price of sandbox land can climb by 5-10 times in less than a year.

Nothing, however, can be guaranteed. This is not financial advice. Do your research.

2. DECENTRALAND (MANA)

Who is Decentraland, and what does it do?

MANA is the ERC20 fungible cryptocurrency token developed by Decentraland. MANA is the currency used in Decentraland's in-meta economy.

The land is bought and sold with MANA in mind.

Decentraland, a virtual world framework, is powered by the Ethereum blockchain. Consumers can produce information and software, experience it, and monetize it. The community owns the property in Decentraland for all time, allowing them to have complete control over the area's development. Users can claim ownership of digital land using a blockchain-based parcel record. Property owners can regulate what information is released on their parcel of land, defined by a set of cartesian coordinate systems (see image below) (x,y). The type of information can range from static 3D scenes to interactive systems such as video games and simulations.

For many investors, the act of purchasing land in Decentraland represents a significant step forward. It is a fantastic technique to increase the amount of money you have. In the same way that real estate is unique, each property portion is represented by a non-fungible token (NFT, ERC 721), which means it cannot be crafted or recreated, much like real estate. In addition, Decentraland offers the option of obtaining a mortgage on the land.

Anyone, at any time, can buy, sell, or lease land peer-to-peer on the official Decentraland Global market or through Opensea. You formally and indisputably own that plot of land when you hold that land token, which is possible because all operations are processed on the Ethereum platform as honest proof-of-purchase. If you decide to invest and build in Decentraland, you should know the following information.

Renting an offline property is expensive, and Decentraland offers a more cost-effective way to own a piece of real estate. The following options are available to those who wish to invest in this platform:

- Purchase a piece of real estate
- Landscape design
- Shop
- Start your own business.
- Play video games set in the real world.
- Socialize and converse with others.

What Can You Buy at The Decentraland Marketplace? What Can You Do In Decentraland?

In the Marketplace, you'll find everything you need to trade and manage your Decentraland tokens.

Land parcels, estates and clothing, and distinctive brands are all available for purchase globally. You can set your MANA price as well as a deadline for submitting proposals.

On Decentraland, you can purchase real estate tracts and estates, as well as wearables and one-of-a-kind names that are for sale.

How To Purchase Land in Decentraland?

Anybody can buy, sell, and lease land on the Decentraland Marketplace or Opensea at any given time. In addition, Decentraland offers the possibility of securing a mortgage on the land itself. In Decentraland, to be precise.

You can obtain a bird's-eye view of every color-coded property/plot, estate, street, region, and plaza in the Decentraland Marketplace using the Atlas View. You may move about the map by clicking and dragging it using the mouse. You can also zoom in and out and hover your mouse pointer over a parcel to display its x, y coordinates, and owner. Any parcels currently available for buy on the global market will be highlighted in this website section. Tap on a parcel to learn more about it, including its status, coordinates, and the public address of its owner (if it has an owner). You could also make a buy or an offer on the displayed parcel from this screen.

What Is an Estate On Decentraland?

An estate, like land, is a digital asset that cannot be exchanged. An estate is a collection of two or more pieces of land located close to one another. These parcels must be adjacent to one another and cannot be separated by a road, plaza, or any other parcel of real estate. You might manage your greater huge estates more effectively by connecting parcels to develop Estates. For example, estates are handy for creating larger sequences that span numerous pieces of land.

What Exactly Is a Parcel, Anyway?

Like real property, every land parcel in Decentraland is a non-fungible token (NFT, ERC 721), which means that it is unique and cannot be solidified or recreated, exactly like it is in a cryptocurrency.

The cheapest piece of land in Decentraland is 3487 MANA.

Where To Buy Items on Decentraland?

The Explore tab will take you to the Marketplace View, where you can see all of the currency for sale. Select the Category option if you just want to look at a specific type of item.

Sort them according to various criteria, such as the most recent, the least priced, and so on.

To keep an eye on products that aren't for sale, turn off the sale feature.

Sorting the products by title will help you find what you're looking for.

How To Buy MANA Tokens

This is a straightforward procedure. Once you've logged into your existing account, click on the "Exchange" or "Markets" link to access the trading platform. Then browse for currency pairs of interest to you, such as ETH/MANA or BTC/MANA, among others. A "BUY" button will appear below, where you can enter the amount of money you wish to spend or even the amount of MANA you want to purchase to proceed. You can purchase MANA tokens by filling out the form on this page.

The following items can be purchased in Decentraland: land, estates, wearables, and one-of-a-kind titles, among others.

Land Prices in Decentraland

The total land area is 90,601 hectares, with 43689 private property parcels, 33886 district land, 9438 roadways, and 3588 plazas. Private land parcels account for 33886 hectares.

Each plot of land has a square area of 16m x 16m, whereas earlier, it was 10m x 10m

The most expensive piece of land ever sold was for 2,000,000 MANA.

The average price of land has increased from less than 500 USD to more than 3000 USD in the last five years alone.

Is It Worthwhile to Invest in Decentraland Real Estate?

The future of Decentraland is based on the number of people who show up and subsequently use the platform. Many people are drawn to this cryptocurrency as a pure digital interest and hope that it will become a legitimate digital currency, like Bitcoin. One of the most major advantages of Decentraland is that it allows users complete ownership access to their virtual assets and properties. This feature distinguishes it from other virtual reality systems on the market.

The money generated and the people who use their land are retained by the proprietors of the digital world's real estate properties. This differs from other systems because it involves a cut of profits. Because the plan is decentralized, there is no centralized authority to oversee or control it in the traditional sense. As a result, purchasing virtual land in the Decentraland Metaverse is a fantastic long-term investment.

Purchasing and selling land on Decentraland entails a certain amount of financial risk.

There is, in fact, a monetary danger. So please exercise caution. In the long run, the value of land may decline. It is still in its early stages and has not gained widespread adoption. Nonetheless, given the current trends, it is doubtful that we will see a significant decline in land prices in the near future. Now would be a wonderful moment to invest if there is a market for it. Decentraland is undoubtedly one of the top coins to look out for as the metaverse develops.

3. STAR ATLAS

In your opinion, when was the last time a AAA game made its roadmap public, released a game iteratively, and engaged in blockchain-native sales? Right, never. What is Star Atlas?

With its open-world, space exploration, and grand strategy elements set in an alternate universe, Star Atlas is an MMORPG that emphasizes player ownership and play-and-earn features in its galactic setting. Players pilot ships and participate in trade, commerce, and combat with other players over limited resources. The company wants to construct a AAA game using the Solana blockchain. This tier 1 blockchain system can handle 50,000+ transactions per second. This will be critical in dealing with the large quantities of transactions in-game, where players will mine and explore, pay taxes, and trade assets — all of which will be powered by Solana and settled on Serum, a decentralized exchange — and will be handled by Solana.

In its biggest ambition, the game strives to be:

- A space-fantasy RPG featuring a real-money economy and NFT assets.

- A grand strategy game involving politics, trading routes, and economically productive territory.
- A 24/7 virtual economy where users may come together in real-time to trade, make contracts, and participate in the battle.

It is possible to explore a completely realistic 3D world with film-quality graphics, driven by Unreal Engine's Nanite, in virtual reality.

Star Atlas has been profiting on the tremendously hot fundraising market for blockchain games. In addition, they announced Animoca Brands as one of its stakeholders, hosted town halls, and published a blockbuster teaser for the highly anticipated game. This corresponds to the first round of their non-fungible token (NFT) pre-sale as well as the launch of their Galactic Asset Offering (GAO) (GAO).

Furthermore, a significant part of the attractiveness of Star Atlas is the opportunity for players to feel like they have ownership in a video game while also being able to monetize their skills, enthusiasm, and effort. Star Atlas is enticing to many players who dream of making a living inside an immersive video game universe. A player can make money by stealing from traders and then selling their stolen goods for $ATLAS (Star Atlas' token) on the marketplace and exchanging that $ATLAS for fiat money.

Key Features

Territory And Exploration

Star Atlas is a game about space and spatial exploration. Initially, players will start at a corner of the map. They will survey the visible stars for celestial and earthly valuables. These basic assets can be refined and exchanged, and sold, just like any other commodity. Those who travel towards the center of the map are rewarded, but with reward comes risk: players may lose their hard-earned riches if they don't act quickly enough. Traveling and transporting things in Star Atlas takes time, which opens new markets for logistics, freight, and even infrastructure, such as bridges, as a result.

A Virtual Economy and Society That Is Strong and Resilient

Among the features planned for Star Atlas is a complex virtual economy in which players can make economic decisions about cargo shipping, travel, fuel management, and defense powering. Player earnings can be obtained in-game through various roles, according to their whitepaper. Here's a sample of what you can expect: "CEO, Bounty Hunter, Repair, Freight, Rescue, Refiners, Miners, Managers'. "Managers are accountable for ensuring that resources are utilized efficiently to create value and utility," according to the authors of an economics article on the metaverse. Power Plant Manager and Salvage Operator are two examples of management-type jobs."

Star Atlas Cryptocurrency Tokens

$ATLAS and $POLIS are the two primary currencies offered by Star Atlas. $ATLAS is an inflatable currency intended for selling in-game assets. At the same time, $POLIS is a low-velocity store of value with a fixed supply (we've seen this model before in Axie Infinity's $SLP and $AXS). While $ATLAS is acquired (and spent) within the game through fighting and exploration, $POLIS empowers players to own and manage space cities. It also reflects a financial investment in the game and governance powers over in-game topics like land tax rates.

Star Atlas aspires to be a physical embodiment of the blockchain, not just in function but also in appearance. In its whitepaper, Star Atlas defines the fundamental mechanic of the game: mining. You stake out celestial and earthly properties claims, and then you start mining the resources. This is accomplished using an NFT ship. A decentralized exchange (Serum's on-chain order book) is used for trading any assets you produce. This allows Star Atlas to support its development by potentially taking a share of secondary sales while also protecting the uniqueness of player assets by assuring that no assets can ever be replicated or destroyed.

Star Atlas' Future

The latest NFT ship sale by Star Atlas may have netted the company more than $20,000,000, but there is still a lot of work to be done. The blockchain will force Star Atlas to maintain grey markets open. In contrast, Star Citizen secured many funding rounds over several years and deliberately clamped down on them. This means that any player dissatisfied with the game can quit and take their money with them. They intend to contribute the USD collected from their ship sales to an ATLAS: USDC automated market maker, which is now under construction. As a result, this money is available to gamers who wish to exchange their in-game currency for other digital goods.

Additionally, Star Atlas has received all-star backing from various sources, including Animoca Brands, Serum, and Moonwhale Ventures. They will provide their DeFi and NFT gaming expertise to help take Star Atlas to the highest level possible.

Moreover, they have a significant AAA gaming partner in Sperasoft, who has collaborated on notable titles such as Star Wars: The Old Republic, Star Wars: Battlefront II, and most recently, Halo Infinite, among others. Being able to rely on a partner with AAA development experience will be critical for Star Atlas' long road ahead. Their recent increase may help them devote additional resources to developing the Star Atlas universe.

In the end, Star Atlas intends to release its game in parts, with the first of them beginning on Monday with the first round of their Galactic Asset Offering (GAO), in which they auctioned ships on their marketplace, which Serum powers.

Star Atlas Roadmap

The CEO of Star Atlas, Michael Wagner, commented that as from November 2021, players and members of the STAR ATLAS community could begin earning from next month by engaging in the second phase: a web-minigame that allows users to use some of their assets and that an immersive 3D environment will be available to explore by the end of the year.

The gamers will begin familiarizing themselves with the game mechanisms and remain involved with the game when released. Typically, AAA games take too long to receive input from their players and the gaming community. Eight million pre-orders for Cyberpunk 2077 were based only on anticipation and trailers before players had a chance to see any of the gameplay elements.

Star Atlas's approach surrounding release flips the conventional secretive development process on its head: offering public information about gameplay mechanisms, marketing its pre-sales, and facilitating a 58,200-person large Discord group. This has its perks. Players can begin immersing themselves in the universe, taking up roles in the Discord (Mercenary, Bounty Hunter, CEO...), and forging alliances and guilds. They can even seek support onboarding into crypto from other prospective players and keep each other interested in the game.

Clicking Play Now on their landing page sends you to their marketplace. The founders have stated that their economy and speculating and trading will be a significant game component. However, it may be a concerning indicator if gamers are more interested in flipping ships than flying them. Moreover, conventional gamers may get disinterested if they are confronted with the prospect of making payments and trading even before they enter the game.

Star Atlas will be the first AAA game to show how NFTs and digital asset ownership can fund game development and operations. By being the dominant liquidity source of $ATLAS and $POLIS, they should support the game's development. This will be a significant step forward in legitimizing play-and-earn as a viable business model for game creators and publishers.

Star Atlas is an enormously intriguing project, but with great potential comes significant risk. If played correctly, Star Atlas can educate game designers on how to employ actual digital scarcity to enhance gamers' experience by offering a play-and-earn metaverse where they can win rewards for their efforts. To see this idea through, their 55-person production team, their partners Hydra Studios and Sperasoft, and their 58,200-member Discord community must work together.

4. ILLUVIUM

AAA blockchain games seem to be the focus for metaverse crypto coins. Illuvium is one such token.

Illuvium is a cross between an open-world RPG and an auto battler, with an economy built on collectible NFTs and resource mining as its primary sources of income.

Will Illuvium be the first AAA blockchain game? That is what Kieran and Aaron Warwick, the company's co-founders, seek to do.

A claim like this has been made numerous times in the brief history of non-traditional gaming (NFT). On the other hand, Illuvium has acquired a substantial pace since 2020.

In just six months, the game reached the milestone of 100,000 Discord subscribers. In addition, between July 2021 and October 2021, the ILV token's value increased from $30 to $700, thanks to the token's introduction on major exchanges such as CoinSpot.

The gameplay experience promised by Illuvium and the game's tokenomics is present in even greater detail than usual.

Is Illuvium possible to be one of the finest NFT games of 2021?

Illuvium is slated to enter open beta in the first quarter of 2022. However, that is already a delay from the third quarter of 2021. As far as we can tell from the limited video available as of September 2021, the Q1 2022 release date appears unrealistic.

Not every region presents the same level of challenge or even accessibility. There is a free-to-play foundation Tier 0 realm that can be explored by anyone who wishes to do so. It helps you become more familiar with the game's mechanics and the concept of tracking down and collecting Illuvial. You can mine for free Shards, which can capture the most fundamental Illuvials in the game.

However, if you want to advance to the Tier 1 and higher regions (it appears that it goes all the way up to Tier 5), you will have to spend some money on it. You'll make your payment in Ethereum (ETH).

What Media Formats Does Illuvium Support?

Although you can lock in PC and maybe Mac, there have been no announcements regarding formats. Illuvium: Zero is a mobile spin-off game currently in development. The current aim is to mine resources for the main game while on the go.

What Is the Premise of Illuvium?

Illuvium starts with a very familiar adventure-themed premise.

In Illuvium, you role-play a member of the intergalactic space fleet in a drastic position. Your ship has crash-landed on a disaster-ravaged planet. The ocean has swallowed most of the land, and what little land remains is being pummeled by natural calamities. Enormous obelisks, erected by some past residents, have blocked the entrance to some locations totally, and giant obelisks have completely blocked access to others.

Your objective is to figure out what caused this catastrophe to occur while also unlocking the obelisks along the way. For this, you will need to mine the soil for Shards, which can be used to capture Illuvials and subdue them so that they can be employed as soldiers in your private army.

Illuvial are God-like creatures that inhabit this unnamed planet, fueled by radiation and performing miracles. You can think of them as being very similar to the Pokémon franchise. Or, at the very least, Pokémon has been given a mature kicking in the shins, removing some of the corny Nintendo lusters in the process.

At the beginning of the game, you will have the option of customizing your character's appearance. As a bonus, you will have the option of selecting a drone to accompany you on your quest as a sidekick. PSD stands for Polymorphic Subordinate Drone, and it is a type of drone.

Illuvium bills itself as an open-world role-playing game, putting it on par with titles such as The Elder Scrolls or Cyberpunk 2077 in terms of scope. If it was AAA, that is. Early footage reveals a 3D world with huge environments loaded with a reasonable amount of detail and a reasonable amount of detail. It's created with Unreal Engine 4, a middleware solution that is exceptionally robust for this type of game. Furthermore, the color scheme is reminiscent of the most popular game powered by that engine, Fortnite.

Although the founders have promised a variety of locations, however, they have only released a few glimpses of what they might look like so far - mainly in the form of concept art – so far.

Some film of a barren rocky island, surrounded by beaches and ocean, has been posted on the internet. However, it appears bland and linear for something meant to be an open world. In addition, a day-night cycle has been proposed.

Traveling around the overworld is accomplished on foot or by passing through the Obelisks dispersed throughout the globe. Towns can also be found near these Obelisks, which will allow you to construct new equipment and serve as a haven against Illuvial.

As you go over the terrain, you will come across Illuvials, land-based creatures. These aren't free-roaming creatures but rather ones who come and go from the earth. Which is a fancy way of stating they just appear out of nowhere during random bouts. This gameplay mechanic can be found in vintage Pokémon and Final Fantasy games. You won't be able to track them down like you would with Monster Hunter.

You can engage in combat with Illuvial. If you are victorious, you will have the opportunity to capture them in a Shard.

Suppose you're attempting to capture an Illuvial. In that case, your chances of success are determined by two factors, namely: the abilities of your Shard and the power of the Illuvial you're trying to capture. Captured Illuvials become a part of your collection, which you can use in future battles. The better and stronger your group, the further you can proceed through the game's various levels.

What Are Illuvials, And How Do They Work?

The Illuvial is the Pokémon of this world, and they have a lot of power. Alternatively, you could call them the Axis. There are over 100 to collect, which is a tad lacking in quantity. Pokémon Sword and Shield, for example, had 400 points. The Illuvial are classified into one of five affinities (water, earth, fire, nature, and air) and five classes (water, earth, fire, nature, and air) (Fighter, Guardian, Rogue, Psion, and Empath). They also have three life phases to progress through, from cubs to deities, during which they can level up.

However, as you go through the game and encounter rarer and more powerful Illuvials, you will find that they can have various affinities and classes. It's also possible to fuse three fully leveled Illuvials, resulting in new variations with rarer powers. The game also rewards players who amass extensive collections of Illuvials like one another, allowing them to "synergize" and increase their abilities.

Adding an extra curveball to the mix is a rare occurrence in photography. The use of a Shard increases the likelihood that it will mutate into a Shiny, Rainbow, or Holo form, each becoming increasingly rare as time passes. These modifications alter the appearance of the Shard and increase its value in the eyes of the NFT's owner.

This capture procedure appears to have some interesting (albeit not unique) depth on paper, which makes it worth mentioning. While those in search of some uncommon (read: valuable) NFTs have something to look forward to from the Fusion concept, Illuvials will also level up during combat, so they will potentially improve as they are utilized in-game, which is something to look forward to from the Fusion concept.

Every time you capture an Illuvial, an NFT is created in your account. It is possible to develop an Illuvial by joining three NFTs together; however, the three existing NFTs are destroyed. However, it is unclear if the outcome of a Fusion is determined by chance or by design.

What Is the Battle System Like In Illuvium?

It is no coincidence that Illuvium's fighting draws inspiration from the auto battler genre, which is a new gameplay experience that only gained traction a few years ago. Auto Chess was the game that started the trend. Still, Dota Underlords, Hearthstone Battlegrounds, and Teamfight Tactics are all crucial inclusions in the list.

An auto battler is a video game where you do not directly control the actual battling in real-time. Instead, it's all about teamwork and planning of time. Use your Illuvials in this scenario, to be precise. What kind of Illuvials have you amassed? What methods did you use to enhance them and provide them with resources? The numbers, rank, class, and type of your formed team are all critical, but how do they interact with one another? Each Illuvial has a fundamental, critical, and final attack that must be considered as well.

Then, of course, these techniques must be evaluated considering your opponent's advantages and disadvantages as well.

Illuvium Vs. Pokemon.

In this sense, the gameplay is highly like that of a card collecting game (CCG). Alternatively, it might be a Pokémon title. On the other hand, with an auto battler, after the pieces (read: Illuvials) are placed on the battlefield, you can step back and observe in a passive capacity to determine if your pre-fight strategy was successful. It is not yet clear how many fighters you will be able to bring into each battle, but screenshots indicate that you will get eight Illuvial.

Lastly, it's important to remember that your player character is always on the battlefield. While the player is a spectator in battle, their avatar actively participates. It can add affinity and class auras to your Illuvial team, which will improve the attributes of your assembled group of Illuvial.

Also available is the option to bond your character with an Illuvial, which will grant even more additional bonuses. This hasn't been fully explained yet, but it is described as semi-permanent for the time being. It suggests that while taking this route may allow you to capture more Illuvials because you have a more powerful team, doing so may deflate the value of that NFT or even cause it to be burned as a sacrifice.

Resources Derived from Mining

Your drone is employed to extract minerals from the planet's surface. It is possible to find ore, uncured Shards, and jewels, all of which are of variable rarity. These can be utilized to make new armor and weapons and improve those you already have on hand. The changes you make here affect the auras your player character emits in battle, affecting the boosts that your Illuvials receive.

You can also harvest organic materials from the trees found all around the earth. These can be given directly to Illuvials to provide limited-time enhancements while in combat. One such example allows you to gather toxic goo, which may be used to boost the attack stats of Illuvial.

What Kinds of Game Modes Are There?

At the game's debut, there will be only one mode available: Adventure Mode. A large portion of the gameplay described above takes place in this mode. We may expect two battle arena modes to be released following the game's introduction: Ranked and Leviathan. The former levels the playing field for skill-based matches, whereas the latter does the opposite. The latter is a free-for-all where you can bring whatever collection you want to share.

It's worth noting that the creator has stated that he intends to allow for in-game betting on the battle arena and arena mode. It will be interesting to see how governments worldwide react in response to this!

What Are Shards?

As you've no doubt guessed, Shards are the Illuvium equivalent of the Pokéball. It is necessary to mine these from the ground in uncured form. How The quality of the Shard you recover is random in terms of strength – or, to put it another way, random in that the stronger the Shard, the more uncommon it is. To capture powerful Illuvials, you'll need powerful Shards. As a result, if you venture out without a few decent Shards up your sleeve, you face the possibility of encountering an Illuvial that you are unable to capture.

Who Is Behind the Development Of Illuvium?

Unfortunately, the developer of Illuvium has chosen to remain anonymous, as has been the case with many other developers in the NFT and blockchain gaming field. Moreover, this is usually a red indicator in my book. Further worries are raised while looking at the two co-founders. Kieran and Aaron Warwick are brothers from Sydney, Australia, who have no previous experience in the game production industry.

Kieran, the first, is a successful entrepreneur. Aaron has never worked in a game development environment despite his many years of experience as a long-time coder enthusiast. The greater crew has a combined total of relatively little noteworthy gaming experience. Nate Wells, a former employee of Irrational Games (BioShock), Crystal Dynamics (Tomb Raider), and Arkane Studios (Dishonored), is the only notable name on the list. He only serves in an advisory/producer capacity.

This does not rule out the possibility of them creating a fantastic game. Still, one should be skeptical of any first-time developer who claims to revolutionize NFT gaming by releasing a AAA title.

The game is being developed on the popular Unreal Engine 4 middleware solution. The tokenomics are all anchored by the Ethereum ERC-20 blockchain, just to be clear. Immutable X oversees dealing with non-financial trading (NFT).

How To Make Money on Illuvium

Because this is non-financial-transaction gaming, the primary means of making money in Illuvium is collecting and selling high-value Illuvial. However, for the player to access the most valuable and rare Illuvials, they must delve deep into the game's world.

To accomplish this, players must mine rirrf3ri Shards from the world's land, unearthing ore, and gemstones that can be used to upgrade and improve gear and harvest organic resources from plants that can provide in-battle buffs, among other things. All of this happens while you're capturing, leveling, and fusing an Illuvial team that's capable of taking on and capturing the rarest (read: most valuable) creatures you can find.

Accordingly, players can trade resources they have gathered, weapons and gear that they have crafted, or lesser Illuvials to other players in markets that have been set up for this purpose. In-game transactions between players are carried out entirely using the ILV token. IlluviDEX is the marketplace name where all these transactions occur, and Immutable X. operates it. This is a third-party, layer-2 Ethereum transaction engine that allows for the transfer of NFTs without any gas.

Furthermore, it should be noted that there is a limited amount of Illuvials in the game's ecological system. As more of a type is discovered, the more difficult it becomes to locate them.

However, there are plans to include more areas and Illuvial in the future. It will be fascinating to watch how they handle the addition of new holders without depreciating or aggravating the existing holders.

Even while it will not be available at launch, gamers will eventually place wagers on ranking matches in the battle arenas. However, it appears that land purchases are related to the Illuvium: Zero mobile spin-off game, as indicated by the fact that they are listed as "coming soon" on the website.

How Much Does It Cost to Participate In Illuvium's Events?

While in-game trades between players are conducted using ILV tokens, all transactions between the game and its players are conducted over the Ethereum blockchain. You will never require ILV to participate in the game.

There are two methods in which Illuvium can take Ethereum away from a player. The first is that they charge a 5 percent fee for any transaction between two or more players. (An additional 0.5 percent fee is charged, split between Immutable X and the user.)

In-game items can be obtained through the purchase of in-game items. The following are the five things you can purchase with ILV tokens:

Shard Curing

Shard Curing is a technique for transforming mined Shards into Illuvial-capturing Shards.

Travel

Using an obelisk to travel between different locations.

Crafting

If you want to get better equipment quickly, this is the way to go.

Cosmetics

If glitter and looks are your things, go for it.

Revival

If you don't want to wait for a wounded Illuvial to mend, you may use ETH to speed up the procedure. This tactic harkens back to the dreadful monetization methods used in mobile games.

It's unclear at this point how much it will cost in ILV to cure a Shard, but this is at the core of the game's monetization strategy. Without Shards, you will be unable to capture any Illuvial.

Illuvium's Tokenomics

At total dilution, the maximum number of ILV tokens traded on the market will be 10,000,000. Before the game's debut, 9,000,000 coins will be in circulation, with the final million coins being distributed as in-game incentives and for winning tournament matches. The DAO already has 1,500,000 more tokens in its treasury, which will be used to reward players for completing in-game objectives and participating in tournaments.

There are 2,000,000 ILV tokens delivered during seeding, which will be unlocked in March 2022 and then drip-fed out at a rate of one-twelfth of a percent every month for the next year. It is also planned to drip-feed the 1,500,000 ILV tokens held by the team at that time but a pace of 1/36 every month for the next three years.

Illuvium Tokenomics is a new kind of economics. According to the time of this writing, another 3,000,000 tokens are currently being locked away in yield farming for 12-month vesting stints (June 2017).

Those who purchase ILV have the option of staking them in two locations. On the one hand, the direct ILV pool is expected to earn an annual percentage yield of 85 percent (APY). Alternatively, there is a secondary LV/ETH Sushi Liquidity Pool on Sushi.com – where you must stake both ETH and ILV 1:1 – where you can invest more money. This results in an annual percentage yield of 600 percent on the combined sum.

However, because of the opportunity for token owners to claim their share early by converting ILV to sILV, the ultimate pool of 10,000,000 ILV tokens may never be reached. In-game spending of this synthetic alternate token, which has a 1:1 value to the main token, allows individuals who have staked their ILV to do so before their funds are unlocked. Any ILV that was used in these transactions was destroyed.

Holders of ILV will get a full refund of all earnings generated through the Illuvium game, including interest.

The Illuvium Distribution System

Illuvium is a DAO.

The Illuminati Council is elected every three months, with five community members being elected to the positions from among those who have put themselves forward. The square root of an individual's ILV stake is used to determine their voting power. As a result, one ILV equals one vote, 36 ILV equals six votes, and so on. This is done to avoid over-dominance by whales and seeders.

The council has the authority to "discuss and distill technical modifications" and consider dividing the 1,500,000 ILV currently stored in the treasury. Any changes to the protocol, known as

Illuvium Improvement Proposals, must be approved by a supermajority to be implemented.

Where Can You Buy Iluvium?

You can currently purchase Illuvium on the following cryptocurrency exchanges:

CoinSpot

CoinSpot is an Australian cryptocurrency exchange that makes it simple to buy, sell, and trade more than 290 different cryptocurrencies.

Binance

This cryptocurrency exchange was founded in 2014. Binance is the largest cryptocurrency exchange globally in terms of the trading volume. Get started with zero-fee AUD deposits and withdrawals in Australia. Take advantage of minimal trading costs, a diverse range of cryptocurrencies, and local customer service available 24 hours a day.

Cointree

This cryptocurrency exchange operates in the United States. Other exchanges where you can buy Illuvium include:

- KuCoin
- Crypto.com

- Gate.io
- OKEx
- Bithumb Cryptocurrency
- Hotbit Cryptocurrency Exchange

What Is Illuvium: Zero, And How Does It Work?

Illuvium: Zero is a spin-off title to the Illuvium series that will be released on mobile devices in 2022. It's a city-builder, where players can buy a plot of land and then build out a civilization that can mine minerals.

If you are playing Illuvium: Zero for free, those resources are tied to the Illuvium: Zero game and cannot be used elsewhere. However, if you have purchased the game, the resources you mine can be transferred to the main game and used as fuel to aid you in your pursuit of Illuvials in the game's primary mode. Also available is the ability to scan Illuvials who roam your property and use the information gathered to make skins that can be sold to the public as blueprint NFTs.

Illuvium appears to be a promising prospect on paper. When it comes to the collectible Illuvials, there is a clear progression and rarity tree and a clear plan for a developing economy. The combination of 3D exploration and auto battler combat should be particularly effective in Unreal Engine 4.

However, there's very little game development experience in the team and no serious footage of gameplay in action to back up the proposed expertise.

As always, whatever you decide to do, invest wisely as this is not financial advice.

5. $UFO

WHAT IS UFO token? Social gaming coin UFO Gaming is decentralized and may be used on any platform. P2E (Play to Earn) Metaverse, Virtual Land, NFT, Gaming, and IDO Launchpad.
The $UFO token will be needed for all $UFO-related activities. For any interaction with the ecosystem, this is required.

The token may be used in three ways: $UFO, UAP, and Plasma Points.

If you stake $UFO or $UFO-ETH in the Cosmos, you'll get Plasma points, which may be redeemed into UFOep.

Our initial game, 'Super Galactic,' requires Origin UFOep to run.

- UAP is needed to buy, trade, and fuse (breed) NFTs in the game. Playing Super Galactic is the only way to get this item.

To access some of the most anticipated gaming projects, you must stake your UFO tokens or purchase property on a dedicated planet.

There are several chains in our Dark Metaverse. Several chains will be used. UFO's games will cover various genres and niches, launching on some of the most respected chains.

Interoperability across games on the same planet. On June 30th, 2021, UFO was made available for trading. It has an undetermined total supply.

Coins like Bitcoin and Ethereum may be acquired using fiat money on crypto exchanges, but not UFO. In this chapter, we'll show you how to buy UFO by first purchasing Ethereum on any fiat-to-crypto exchange and then transferring your funds to the exchange that trades this currency.

First, you'll need to get your hands on one of the most popular cryptocurrencies, like Ethereum (ETH).

The second step is to buy ETH using a currency other than ETH.

Third, transfer ETH to an altcoin trading platform like Bitfinex.

UFO may be traded on various listed exchanges, so visit each and sign up for an account.

Below are some examples of exchanges that currently possess the UFO token while writing this book.

- Gate.io

- MeXC
- Coinbase
- Binance
- 1-inch
- ShibaSwap
- Uniswap between UFO and WETH (V2)
- UFO/WETH

In addition to the exchanges listed above, there are several well-known cryptocurrency exchanges with large user bases and good daily trading volumes. As a result, you'll be able to sell your coins whenever you choose and at a lesser charge. As a result, if UFO is listed on one of these exchanges, a significant amount of trading volume will be generated by its customers, providing you with excellent trading chances!

Binance

Binance was founded in China, a prominent cryptocurrency exchange, but later relocated to Malta, a crypto-friendly EU island. The crypto-to-crypto exchange services provided by Binance are well-known. After bursting onto the scene during the 2017 crypto frenzy, Binance has become the world's most popular cryptocurrency exchange. Binance does not accept US investors.

Gate.io

Gate.io is a cryptocurrency exchange based in the United States founded in 2017. If you are a US investor, you may trade on this exchange since it is based in the United States. Both English and Chinese versions of the conversation are accessible (the latter being very helpful for Chinese investors). Gate.io's wide range of trading pairs is a key selling point. Most of the most recent cryptocurrencies are available right here. The trade volume on Gate.io is also impressive. It ranks among the top 20 most active stock exchanges in terms of trading volume almost every day. The daily trade volume is about $100 million. The most popular trading pairs on Gate.io often include USDT (Tether) as a component. This exchange's enormous number of trading pairs and outstanding liquidity, to recap, are two of its most striking features.

How To Store $UFO

Storage of UFO in hardware wallets is the last step.

Even though Binance is one of the safest cryptocurrency exchanges, there have been hacking events, stolen assets. If you intend to store your UFO for a long time, you may want to look at protecting it. Due to the nature of the exchange, wallets, which we refer to as "Hot Wallets," will constantly be online, exposing many risks. By far, the most secure method of keeping your coins is to use "Cold wallets," which only allow access to the blockchain (or simply "go online") when sending cash. This reduces the likelihood of a hacking event occurring. An offline-created combination of public and private addresses known as a "paper wallet" is a free cold wallet. You'll be able to write it down someplace and keep it safe. However, it is not a long-term solution and is vulnerable to various dangers.

Hardware wallets outclass cold wallets in this scenario. A USB-enabled gadget is frequently used to save your wallet's most crucial information in a more long-lasting manner. They have military-grade security built-in, and their software is updated regularly by the devices' makers, ensuring their continued safety. The most popular alternatives in this category are the Ledger Nano S and Ledger Nano X, which range in price from $50 to $100 depending on the capabilities they provide. In our view, these wallets are a wise purchase if you want to keep your money safe.

Is UFO A Good Investment?

Its market cap is still deemed tiny, which means that the price of UFO may be quite volatile when the market moves significantly, as it has increased by 949.67 percent over the previous three months. If UFO keeps growing steadily, it will likely see some significant increases shortly. The key to successful trading is never to lose sight of the big picture.

Remember that this is not financial advice.

Investors in cryptocurrency should do their due diligence and use extreme caution.

Conclusion

The Metaverse remains a wholly new concept for technology and the rest of the world. Many regulators, innovators, and commerce merchants are rapidly in the race to understand and begin to ply their goods on the metaverse. While it is still very unclear where deliberations and disruptions of the metaverse might end, one thing is clear: the metaverse is here to stay.

The question to ask is: Are you willing to swim along with the tide?

Printed in Great Britain
by Amazon